PROBLEMS & SOLUTIONS IN THEORETICAL & MATHEMATICAL PHYSICS

Problems and Solutions in Theoretical and Mathematical Physics

Volume I:
Introductory Level

3nd extended and revised edition

by
Willi-Hans Steeb
International School for Scientific Computing

in collaboration with
Yorick Hardy
International School for Scientific Computing

PROBLEMS & SOLUTIONS IN THEORETICAL & MATHEMATICAL PHYSICS

Third Edition

Volume I: Introductory Level

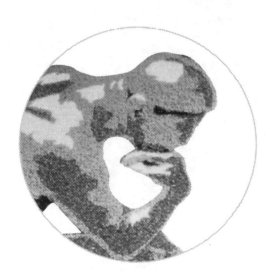

Willi-Hans Steeb
University of Johannesburg
South Africa

World Scientific

NEW JERSEY · LONDON · SINGAPORE · BEIJING · SHANGHAI · HONG KONG · TAIPEI · CHENNAI

Published by

World Scientific Publishing Co. Pte. Ltd.

5 Toh Tuck Link, Singapore 596224

USA office: 27 Warren Street, Suite 401-402, Hackensack, NJ 07601

UK office: 57 Shelton Street, Covent Garden, London WC2H 9HE

British Library Cataloguing-in-Publication Data

A catalogue record for this book is available from the British Library.

PROBLEMS AND SOLUTIONS IN THEORETICAL AND MATHEMATICAL PHYSICS (3rd Edition)
Volume I: Introductory Level

ISBN-13 978-981-4282-14-7
ISBN-10 981-4282-14-6
ISBN-13 978-981-4282-15-4 (pbk)
ISBN-10 981-4282-15-4 (pbk)

Printed in Singapore.

Preface

The purpose of this book is to supply a collection of problems together with their detailed solution which will prove to be valuable to students as well as to research workers in the fields of mathematics, physics, engineering and other sciences. The topics range in difficulty from elementary to advanced. Almost all problems are solved in detail and most of the problems are self-contained. All relevant definitions are given. Students can learn important principles and strategies required for problem solving. Teachers will also find this text useful as a supplement, since important concepts and techniques are developed in the problems. The material was tested in my lectures given around the world.

The book is divided into two volumes. Volume I presents the introductory problems for undergraduate and advanced undergraduate students. In volume II the more advanced problems together with their detailed solutions are collected, to meet the needs of graduate students and researchers. Problems included cover most of the new fields in theoretical and mathematical physics such as Lax representation, Bäcklund transformation, soliton equations, Hilbert space theory, Lie algebra valued differential forms, Hirota technique, Painlevé test, the Bethe ansatz, the Yang-Baxter relation, wavelets, chaos, fractals, complexity, etc.

In the reference section some other books are listed having useful problems for students in theoretical physics and mathematical physics. Some or related problems to those given in volume I can be found in these books. In volume II references are given to books and original articles where some of the advanced problems can be found.

I wish to express my gratitude to Dr. Yorick Hardy for a critical reading of the manuscripts.

Any useful suggestions and comments are welcome.

Email addresses of the author:

steebwilli@gmail.com

steeb_wh@yahoo.com

Home page of the author:

http://issc.uj.ac.za

Contents

Notation

$\emptyset$	empty set
$A \subset B$	subset A of set B
$A \cap B$	the intersection of the sets A and B
$A \cup B$	the union of the sets A and B
$\mathbb{N}$	natural numbers
$\mathbb{Z}$	integers
$\mathbb{Q}$	rational numbers
$\mathbb{R}$	real numbers
$\mathbb{R}^+$	nonnegative real numbers
$\mathbb{C}$	complex numbers
$\mathbb{R}^n$	n-dimensional Euclidian space
$\mathbb{C}^n$	n-dimensional complex linear space
i	$:= \sqrt{-1}$
$\Re z$	real part of the complex number z
$\Im z$	imaginary part of the complex number z
$\mathbf{x} \in \mathbb{R}^n$	element $\mathbf{x}$ of $\mathbf{R}^n$ (column vector)
$\mathbf{0}$	zero vector (column vector)
$f \circ g$	composition of two mappings $(f \circ g)(x) = f(g(x))$
u	dependent variable
t	independent variable (time variable)
x	independent variable (space variable)
$\mathbf{x}^T = (x_1, x_2, \ldots, x_n)$	vector of independent variables, ^{T} means transpose
$\mathbf{u}^T = (u_1, u_2, \ldots, u_n)$	vector of dependent variables, ^{T} means transpose
$\|\cdot\|$	norm
$\mathbf{x} \cdot \mathbf{y} \equiv \mathbf{x}^T \mathbf{y}$	scalar product (inner product) in vector space $\mathbb{R}^n$
$\mathbf{x} \times \mathbf{y}$	vector product in vector space $\mathbb{R}^3$
det	determinant of a square matrix

tr	trace of a square matrix
0_n	$n \times n$ zero matrix
I_n	$n \times n$ unit matrix (identity matrix)
I	identity operator
$[\,,]$	commutator
$[\,,]_+$	anticommutator
δ_{jk}	Kronecker delta with $\delta_{jk} = 1$ for $j = k$ and $\delta_{jk} = 0$ for $j \neq k$
$\epsilon_{jk\ell}$	total antisymmetric tensor $\epsilon_{123} = 1$
$\mathrm{sgn}(x)$	the sign of x, 1 if $x > 0$, -1 if $x < 0$, 0 if $x = 0$
λ	eigenvalue
ϵ	real parameter
$\otimes$	Kronecker product, tensor product
$\wedge$	Grassmann product (exterior product, wedge product)
L	Lagrange function
H	Hamilton function
$\hat{H}$	Hamilton operator
$\mathcal{H}$	Hilbert space
$L_2(\Omega)$	Hilbert space of square integrable functions
$\langle\,,\rangle$	scalar product in Hilbert space
δ	delta function

Chapter 1

Complex Numbers and Functions

A complex number z can be written as

$$z = x + iy, \qquad x, y \in \mathbb{R}$$

where x is the real part $\Re(z)$, y is the imaginary part $\Im(z)$ and $i = \sqrt{-1}$. The number $\bar{z} = x - iy$ is known as the *complex conjugate* of z. A complex number z also admits the polar coordinate representation

$$z = |z|e^{i\phi} = |z|(\cos(\phi) + i\sin(\phi)).$$

The length, r, of the segment $(0,0) - (x,y)$ is known as the absolute value or modulus of z and is denoted by $|z|$. We have $|z| = \sqrt{z\bar{z}}$. The angle which the segment Oz makes with the Ox axis is known as the *argument* (amplitude or angle) of z and is denoted by $\arg z$. We have $\tan(\arg z) = \tan \phi = y/x$ and $\arg z$ is defined as a real number modulo 2π provided $z \neq 0$. Some authors take $0 \leq \arg z < 2\pi$ while others opt for the range $-\pi < \arg z \leq \pi$.

Problem 1. (i) Let $i := \sqrt{-1}$. Calculate

$$i^i.$$

(ii) Let $z = x + iy$ with $x, y \in \mathbb{R}$. Calculate

$$|e^{iz}|.$$

Solution 1.   (i) Let z_1 and z_2 be complex numbers. Assume that $z_1 \neq 0$. One defines

$$z_1^{z_2} := e^{z_2 \ln z_1}.$$

Now

$$\ln z = \ln r + i(\theta + 2k\pi), \qquad k \in \mathbb{Z}$$

where $z = x + iy$, $x, y \in \mathbb{R}$ and

$$r := \sqrt{x^2 + y^2}.$$

Therefore $\ln z$ is an infinitely many valued function. The *principal branch* of $\ln z$ is defined as $\ln r + i\theta$ where $0 \leq \theta < 2\pi$. Consequently, for the principal branch we have

$$i^i = e^{i \ln i} = e^{ii\pi/2} = e^{-\pi/2}.$$

(ii) Since $z = x + iy$ with $x, y \in \mathbb{R}$ the complex conjugate is given by $\bar{z} = x - iy$. Therefore

$$|e^{iz}| := \sqrt{e^{iz} e^{-i\bar{z}}} = \sqrt{e^{i(x+iy)} e^{-i(x-iy)}} = \sqrt{e^{-2y}} = e^{-y}.$$

Problem 2.   Let $z = x + iy$, where $x, y \in \mathbb{R}$. Find the real and imaginary part of

$$\cos\left(\frac{z}{2}\right) \overline{\sin\left(\frac{z}{2}\right)}.$$

Solution 2.   Since

$$\cos\left(\frac{z}{2}\right) \equiv \frac{e^{i(x+iy)/2} + e^{-i(x+iy)/2}}{2}$$

and

$$\overline{\sin\left(\frac{z}{2}\right)} \equiv \sin\left(\frac{\bar{z}}{2}\right) \equiv \frac{e^{i(x-iy)/2} - e^{-i(x-iy)/2}}{2i}$$

we obtain

$$\cos\left(\frac{z}{2}\right) \overline{\sin\left(\frac{\bar{z}}{2}\right)} \equiv \frac{e^{ix} - e^{-ix} + e^{y} - e^{-y}}{4i}$$

or

$$\cos\left(\frac{z}{2}\right) \overline{\sin\left(\frac{\bar{z}}{2}\right)} \equiv \frac{1}{2} \sin x + \frac{1}{2i} \sinh y \equiv \frac{1}{2} \sin x - \frac{i}{2} \sinh y.$$

Consequently, the real part is

$$\frac{1}{2} \sin x$$

and the imaginary part is

$$-\frac{1}{2}\sinh y.$$

Problem 3. Let n be an integer. Then

$$e^{1+2n\pi i} = e.$$

If we write

$$(e^{1+2n\pi i})^{1+2n\pi i} = e^{1+2n\pi i} = e$$

and

$$(e^{1+2n\pi i})^{1+2n\pi i} = e^{1+4n\pi i - 4n^2\pi^2} = ee^{-4n^2\pi^2}$$

it follows that

$$e^{-4n^2\pi^2} = 1.$$

Solve this paradox.

Solution 3. Define the *imaginary remainder* $\Im r(z)$ and the *imaginary quotient* $\Im q(z)$ by

$$\Im(z) = \Im r(z) + 2\pi\Im q(z)$$

where $\Im r(z) \in (-\pi, \pi]$ and $\Im q(z) \in \mathbb{Z}$. We have

$$(e^u)^v = e^{uv}e^{-v2\pi i\Im q(u)}$$

since

$$(e^u)^v = e^{\ln(e^u)^v} = e^{(\Re(u)+i\Im r(u))v} = e^{(u-i2\pi\Im q(u))v} = e^{uv}e^{-vi2\pi\Im q(u)}.$$

Thus the expression

$$(e^{1+2n\pi i})^{1+2n\pi i} = e^{1+4n\pi i - 4n^2\pi^2}$$

should be replaced by

$$(e^{1+2n\pi i})^{1+2n\pi i} = e^{1+4n\pi i - 4n^2\pi^2}e^{-(1+2n\pi i)2\pi i\Im q(1+2n\pi i)}$$

$$= e^{1-4n^2\pi^2}e^{-(1+2n\pi i)2\pi in}$$

$$= e^{1-4n^2\pi^2}e^{-2n\pi i + 4n^2\pi^2}$$

$$= e.$$

In problem 1 we considered the domain $[0, 2\pi)$ for the principal branch. In this problem we consider the domain $(-\pi, \pi]$ for the principal branch.

Problem 4.   Show that
$$\cos(2\theta) \equiv \cos^2 \theta - \sin^2 \theta$$
$$\sin(2\theta) \equiv 2 \sin \theta \cos \theta$$
using De Moivre's Formula, where $\theta \in \mathbb{R}$.

Solution 4.   Let n be a positive integer. We set $z = re^{i\phi}$. Then
$$z^n = (re^{i\theta})^n = r^n e^{in\theta} = r^n (\cos(n\theta) + i\sin(n\theta)).$$
If $|z| = r = 1$, then we have
$$e^{in\theta} \equiv \cos(n\theta) + i\sin(n\theta) \equiv (\cos\theta + i\sin\theta)^n.$$
This is know as *De Moivre's theorem*. Now let $n = 2$. Then it follows that
$$\begin{aligned}
\cos(2\theta) + i\sin(2\theta) &= e^{2i\theta} \\
&= (e^{i\theta})^2 \\
&= (\cos\theta + i\sin\theta)^2 \\
&= \cos^2 \theta - \sin^2 \theta + 2i\sin\theta \cos\theta.
\end{aligned}$$

The real and imaginary parts provide the trigonometric relations given above. We can find trigonometric identities for arbitrary n. For example for $n = 3$ we find the identities
$$\cos(3\theta) \equiv 4\cos^3 \theta - 3\cos\theta, \quad \sin(3\theta) \equiv 3\sin\theta - 4\sin^3 \theta$$
by considering the real and imaginary parts.

Problem 5.   Calculate
$$\Im\left(\frac{1+z}{1-z}\right).$$
Set $z = re^{i\phi}$.

Solution 5.   We have
$$\begin{aligned}
\Im\left(\frac{1+z}{1-z}\right) &= \Im\left(\frac{1+z}{1-z}\frac{1-\bar{z}}{1-\bar{z}}\right) \\
&= \Im\left(\frac{1+z-\bar{z}-z\bar{z}}{1-z-\bar{z}+z\bar{z}}\right) \\
&= \Im\left(\frac{1+r(e^{i\phi}-e^{-i\phi})-r^2}{1-r(e^{i\phi}+e^{-i\phi})+r^2}\right) \\
&= \Im\left(\frac{1+2ir\sin\phi-r^2}{1-2r\cos\phi+r^2}\right) \\
&= \frac{2r\sin\phi}{1-2r\cos\phi+r^2}.
\end{aligned}$$

Problem 6.   Let μ, ν be two complex numbers. Assume that
$$|\mu|^2 - |\nu|^2 = 1\,.$$
Find a parametrization.

Solution 6.   A parametrization would be $(r \in \mathbb{R})$
$$\mu = \cosh(r), \qquad \nu = e^{i\phi}\sinh(r)$$
since $\cosh^2(r) - \sinh^2(r) = 1$ and $e^{i\phi}e^{-i\phi} = 1$.

Problem 7.   Let $\alpha \in \mathbb{R}$ and $z_1, z_2 \in \mathbb{C}$. Calculate
$$R = z_1^* z_2 e^{(e^{i\alpha}-1)z_1 z_1^*} e^{(e^{-i\alpha}-1)z_2 z_2^*} + z_1 z_2^* e^{(e^{-i\alpha}-1)z_1 z_1^*} e^{(e^{i\alpha}-1)z_2 z_2^*}$$
using $z_j = r_j e^{i\phi_j}$ and $e^{i\phi_j} = \cos(\phi_j) + i\sin(\phi_j)$. The expression plays a role in quantum optics. Obviously R is real. Consider then the special case $\phi_1 = \phi_2$.

Solution 7.   Using $\cos(-x) = \cos(x)$ and $\sin(-x) = -\sin(x)$ we have
$$R = r_1 r_2 \big(e^{i(\phi_2-\phi_1)}e^{(e^{i\alpha}-1)r_1^2}e^{(e^{-i\alpha}-1)r_2^2} + e^{i(\phi_1-\phi_2)}e^{(e^{-i\alpha}-1)r_1^2}e^{(e^{i\alpha}-1)r_2^2}\big)$$
$$= r_1 r_2 \cos(\phi_2 - \phi_1)\big(e^{(e^{i\alpha}-1)r_1^2+(e^{-i\alpha}-1)r_2^2} + e^{(e^{-i\alpha}-1)r_1^2+(e^{i\alpha}-1)r_2^2}\big)$$
$$+ ir_1 r_2 \sin(\phi_2 - \phi_1)\big(e^{(e^{i\alpha}-1)r_1^2+(e^{-i\alpha}-1)r_2^2} - e^{(e^{-i\alpha}-1)r_1^2+(e^{i\alpha}-1)r_2^2}\big)$$
$$= 2r_1 r_2 \cos(\phi_2 - \phi_1)e^{(\cos\alpha-1)(r_1^2+r_2^2)}\cos((r_1^2 - r_2^2)\sin(\alpha))$$
$$- 2r_1 r_2 \sin(\phi_2 - \phi_1)e^{(\cos\alpha-1)(r_1^2+r_2^2)}\sin((r_1^2 - r_2^2)\sin(\alpha))\,.$$
For the special case $\phi_1 = \phi_2$ we obtain
$$R = 2r_1 r_2 e^{(\cos\alpha-1)(r_1^2+r_2^2)}\cos((r_1^2 - r_2^2)\sin(\alpha))\,.$$

Problem 8.   Show that for all $x, y \in \mathbb{R}$
$$|e^{ix} - e^{iy}| = 2\left|\sin\left(\frac{x-y}{2}\right)\right|\,.$$

Solution 8.   We have
$$|e^{ix} - e^{iy}| = \left|e^{i(x+y)/2}\right| \cdot \left|e^{i(x-y)/2} - e^{-i(x-y)/2}\right|$$
$$= |e^{i(x-y)/2} - e^{-i(x-y)/2}|$$
$$= 2\left|i\sin\left(\frac{x-y}{2}\right)\right|$$
$$= 2\left|\sin\left(\frac{x-y}{2}\right)\right|\,.$$

Problem 9.   Let $x, y \in \mathbb{R}$ and

$$A = \begin{pmatrix} x & -y \\ y & x \end{pmatrix}.$$

Calculate $\exp(A)$. Use the fact that there is a field *isomorphism* between the complex numbers $z = x + iy$ and the 2×2 matrices

$$\begin{pmatrix} x & -y \\ y & x \end{pmatrix}.$$

Solution 9.   We have

$$\begin{aligned}
e^z &= e^{x+iy} \\
&= e^x e^{iy} \\
&= e^x (\cos y + i \sin y) \\
&= e^x \cos y + i e^x \sin y.
\end{aligned}$$

Thus the real part of e^z is given by $e^x \cos y$ and the imaginary part of e^z is given by $e^x \sin y$. Owing to the isomorphism we have

$$e^A = \begin{pmatrix} e^x \cos y & -e^x \sin y \\ e^x \sin y & e^x \cos y \end{pmatrix}.$$

Problem 10.   Given the equation

$$e^x - e^{-x} = -2iy.$$

Show that

$$e^x = -iy \pm \sqrt{1 - y^2}.$$

Solution 10.   Multiplying the equation by e^x yields

$$(e^x)^2 + 2iye^x - 1 = 0.$$

Solving this quadratic equation for e^x provides the result.

Chapter 2

Sums and Products

Arithmetic series, geometric series and the harmonic series play a role in problems in theoretical and mathematical physics. An *arithmetic series* is the sum of a sequence $\{\, s_k \,\}$, $k = 0, 1, \ldots$ in which each term is calculated from the previous one by adding (or subtracting) a constant c. This means

$$s_k = s_{k-1} + c = s_{k-2} + 2c = \cdots = s_0 + c \cdot k$$

where $k = 1, 2, \ldots.$ A *geometric series* is a series with a constant ratio between sucessive terms. The *harmonic series* is the infinite series

$$\sum_{k=1}^{\infty} \frac{1}{k}.$$

The harmonic series diverges to infinity.

Problem 1. (i) Calculate

$$S = \sum_{n=1}^{\infty} \frac{1}{3^n}. \tag{1}$$

(ii) Calculate

$$S(n, x) - \sum_{k=0}^{n} x^k$$

where $x \neq 0$. The two series are so-called *geometric series*.

Solution 1.   (i) We can use the ratio test or the nth root test to prove that the series (1) converges absolutely. We have

$$S = \sum_{n=1}^{\infty} \frac{1}{3^n} \equiv \frac{1}{3} + \frac{1}{9} + \frac{1}{27} + \frac{1}{81} + \cdots \equiv \frac{1}{3}\left(1 + \frac{1}{3} + \frac{1}{9} + \frac{1}{27} + \cdots\right) \equiv \frac{1}{3}(1+S).$$

Consequently, $S = \dfrac{1}{2}$.

(ii) Let $x = 1$, then obviously

$$S(n,1) = \sum_{k=0}^{n} 1 = n+1.$$

Now let $x \neq 1$. Since

$$S(n,x) = \sum_{k=0}^{n} x^k = 1 + x + x^2 + \cdots + x^n$$

we obtain

$$xS(n,x) = \sum_{k=0}^{n} x^{k+1} = x + x^2 + \cdots + x^n + x^{n+1}.$$

Subtracting the two equations gives

$$(x-1)S(n,x) = -1 + x^{n+1}$$

or

$$S(n,x) = \frac{1 - x^{n+1}}{1 - x}.$$

Problem 2.   Let $n \in \mathbb{N}$. Evaluate the sum

$$\sum_{k=1}^{n} \frac{k^2}{2^k}. \tag{1}$$

Solution 2.   If we define

$$S(x) := \sum_{k=1}^{n} k^2 x^k$$

then the sum (1) is given by $S(\frac{1}{2})$. Now we can use the techniques of analysis. We know that

$$\sum_{k=0}^{n} x^k = \frac{1 - x^{n+1}}{1 - x}, \qquad x \neq 1.$$

Differentiating each side with respect to x yields

$$\sum_{k=1}^{n} kx^{k-1} = \frac{1 - (n+1)x^n + nx^{n+1}}{(1-x)^2}.$$

Multiplying each side of this equation by x, differentiating a second time, and multiplying the result by x yields

$$S(x) = \sum_{k=1}^{n} k^2 x^k = \frac{x(1+x) - x^{n+1}(nx - n - 1)^2 - x^{n+2}}{(1-x)^3}.$$

From this equation it follows that

$$S\left(\frac{1}{2}\right) = \sum_{k=1}^{n} \frac{k^2}{2^k}$$

$$= 6 - \frac{1}{2^{n-2}}\left(\frac{1}{2}n - n - 1\right)^2 - \frac{1}{2^{n-1}}$$

$$= 6 - \left(\frac{n^2 + 4n + 6}{2^n}\right).$$

Problem 3. Calculate

$$S(\alpha) = \sum_{n=0}^{N-1} e^{in\alpha} \tag{1}$$

where $\alpha \in \mathbb{R}$ and $N \in \mathbb{N}$.

Solution 3. Let $\alpha = 2m\pi$ where $m \in \mathbb{Z}$. Since

$$e^{2inm\pi} = 1, \quad n \in \mathbb{N} \cup \{0\}$$

we find

$$S(2m\pi) = N.$$

Now let $\alpha \neq 2m\pi$, where $m \in \mathbb{Z}$. Owing to $e^{in\alpha} \equiv (e^{i\alpha})^n$ the sum (1) over n is a geometric series. Consequently,

$$S(\alpha) = \sum_{n=0}^{N-1} e^{in\alpha} = \frac{1 - e^{iN\alpha}}{1 - e^{i\alpha}} = \frac{\sin(N\alpha/2)}{\sin(\alpha/2)} e^{i(N-1)\alpha/2}.$$

The sums

$$\sum_{n=1}^{N-1} ne^{in\alpha}, \qquad \sum_{n=1}^{N-1} n^2 e^{in\alpha}$$

can now easily be calculated since

$$\frac{dS}{d\alpha} = i \sum_{n=1}^{N-1} n e^{in\alpha}$$

and

$$\frac{d^2 S}{d\alpha^2} = - \sum_{n=1}^{N-1} n^2 e^{in\alpha}.$$

Problem 4.   Calculate

$$\sum_{k=-N}^{N} \frac{1}{1 + e^{\lambda \epsilon(k)}} \tag{1}$$

where $N \in \mathbb{N}$, $\epsilon(-k) = -\epsilon(k)$ and λ is a real parameter.

Solution 4.   We define

$$S(N, \lambda) := \sum_{k=-N}^{N} \frac{1}{1 + e^{\lambda \epsilon(k)}}.$$

Taking the derivative of S with respect to λ we obtain

$$\frac{dS}{d\lambda} = - \sum_{k=-N}^{N} \frac{\epsilon(k) e^{\lambda \epsilon(k)}}{(1 + e^{\lambda \epsilon(k)})^2} \equiv - \sum_{k=-N}^{N} \frac{\epsilon(k)}{2 + 2\cosh(\lambda \epsilon(k))}.$$

Since $\epsilon(-k) = -\epsilon(k)$ and $\cosh(x)$ is an even function we find

$$\frac{dS}{d\lambda} = 0.$$

From $S(N, \lambda)$ we obtain the initial condition for the differential equation

$$S(N, \lambda = 0) = \sum_{k=-N}^{N} \frac{1}{1 + 1} = \frac{1}{2}(2N + 1).$$

Consequently, the solution of the linear differential equation (3) with the initial condition (4) is given by

$$S(N, \lambda) = N + \frac{1}{2}.$$

This means the sum (1) is independent of λ. This sum plays a role in solid state physics.

Problem 5. Let

$$z = \sum_{n \in I} a_n e^{-ix_n} \tag{1}$$

where I is a finite set and a_n, $x_n \in \mathbb{R}$. Calculate $z\bar{z}$, where $\bar{z}$ denotes the complex conjugate of z.

Solution 5. From (1) we obtain

$$\bar{z} = \sum_{m \in I} a_m e^{ix_m}.$$

Therefore

$$z\bar{z} = \sum_{n \in I} \sum_{m \in I} a_n a_m e^{-ix_n} e^{ix_m} = \sum_{n \in I} \sum_{m \in I} a_n a_m e^{i(x_m - x_n)}.$$

Thus

$$z\bar{z} = \sum_{n \in I} \sum_{m \in I} a_n a_m (\cos(x_m - x_n) + i \sin(x_m - x_n)).$$

Since $\sin(-\alpha) \equiv -\sin(\alpha)$ we have

$$\sin(x_m - x_n) \equiv \sin(-(x_n - x_m)) \equiv -\sin(x_n - x_m).$$

Therefore

$$\sum_{n \in I} \sum_{m \in I} a_n a_m \sin(x_m - x_n) = 0.$$

Consequently,

$$z\bar{z} = \sum_{n \in I} \sum_{m \in I} a_n a_m \cos(x_m - x_n).$$

Problem 6. Let a be a positive constant. Let

$$S := \left\{ \frac{a}{2}(1,1,1), \ \frac{a}{2}(1,1,-1), \ \frac{a}{2}(1,-1,1), \ \frac{a}{2}(-1,1,1), \right.$$
$$\left. \frac{a}{2}(1,-1,-1), \ \frac{a}{2}(-1,1,-1), \ \frac{a}{2}(-1,-1,1), \ \frac{a}{2}(-1,-1,-1) \right\}.$$

Calculate

$$\sum_{\boldsymbol{\Delta} \in S} e^{i(\mathbf{k} \cdot \boldsymbol{\Delta})}$$

where $\mathbf{k} \cdot \boldsymbol{\Delta} := k_1 \Delta_1 + k_2 \Delta_2 + k_3 \Delta_3$.

Solution 6.

$$\sum_{\mathbf{\Delta}\in S} e^{i(\mathbf{k}\cdot\mathbf{\Delta})} \equiv \sum_{\mathbf{\Delta}\in S} e^{i(k_1\Delta_1+k_2\Delta_2+k_3\Delta_3)}$$

$$\equiv e^{i(k_1a/2+k_2a/2+k_3a/2)} + e^{-i(k_1a/2+k_2a/2+k_3a/2)}$$
$$+ e^{i(k_1a/2+k_2a/2-k_3a/2)} + e^{-i(k_1a/2+k_2a/2-k_3a/2)}$$
$$+ e^{i(k_1a/2-k_2a/2+k_3a/2)} + e^{-i(k_1a/2-k_2a/2+k_3a/2)}$$
$$+ e^{i(-k_1a/2+k_2a/2+k_3a/2)} + e^{-i(-k_1a/2+k_2a/2+k_3a/2)}.$$

Thus

$$\sum_{\mathbf{\Delta}\in S} e^{i\mathbf{k}\cdot\mathbf{\Delta}} \equiv 2[\cos(k_1a/2+k_2a/2+k_3a/2) + \cos(k_1a/2+k_2a/2-k_3a/2)$$
$$+ \cos(k_1a/2-k_2a/2+k_3a/2) + \cos(-k_1a/2+k_2a/2+k_3a/2)].$$

For $b, c, d \in \mathbb{R}$ we have the identity

$$\cos(b+c+d)+\cos(b+c-d)+\cos(b-c+d)+\cos(-b+c+d) \equiv 4\cos(b)\cos(c)\cos(d).$$

Thus we obtain

$$\sum_{\mathbf{\Delta}\in S} e^{i(\mathbf{k}\cdot\mathbf{\Delta})} \equiv 8\cos\left(\frac{k_1 a}{2}\right)\cos\left(\frac{k_2 a}{2}\right)\cos\left(\frac{k_3 a}{2}\right).$$

Problem 7.   (i) Let

$$A_j := \frac{1}{x-a_j}, \qquad B_{kj} := \frac{1}{a_k-a_j} \tag{1}$$

where $k \neq j$. Show that

$$A_n B_{ns} + A_s B_{sn} = A_n A_s. \tag{2}$$

(ii) Show that the identity

$$\sum_{\substack{j,k=1 \\ j\neq k}}^{n} \frac{1}{x-a_j}\frac{1}{x-a_k} \equiv 2\sum_{\substack{j,k=1 \\ j\neq k}}^{n} \frac{1}{x-a_k}\frac{1}{a_k-a_j}$$

holds. Using the identity (2) we can write this equation as

$$\sum_{\substack{j,k=1 \\ j\neq k}}^{n} A_j A_k \equiv 2\sum_{\substack{j,k=1 \\ j\neq k}}^{n} A_k B_{kj}.$$

Solution 7. (i) By straightfoward calculation we find

$$A_n B_{ns} + A_s B_{sn} = \frac{1}{x - a_n} \frac{1}{a_n - a_s} + \frac{1}{x - a_s} \frac{1}{a_s - a_n}$$

$$= \frac{1}{a_n - a_s} \left(\frac{1}{x - a_n} - \frac{1}{x - a_s} \right)$$

$$= \frac{1}{a_n - a_s} \left(\frac{x - a_s - x + a_n}{(x - a_n)(x - a_s)} \right)$$

$$= \frac{1}{(x - a_n)(x - a_s)}$$

$$= A_n A_s.$$

(ii) We find that

$$\sum_{\substack{j,k=1 \\ j \neq k}}^{n} A_j A_k \equiv \sum_{\substack{j,k=1 \\ j \neq k}}^{n-1} A_j A_k + \sum_{s=1}^{n-1} A_n A_s + \sum_{t=1}^{n-1} A_t A_n.$$

Using (2) we obtain

$$\sum_{\substack{j,k=1 \\ j \neq k}}^{n} A_j A_k \equiv 2 \sum_{\substack{j,k=1 \\ j \neq k}}^{n-1} A_k B_{kj} + 2 A_n \sum_{s=1}^{n-1} A_s$$

and

$$A_n \sum_{s=1}^{n-1} A_s \equiv \sum_{s=1}^{n-1} (A_n B_{ns} + A_s B_{sn}) \equiv \sum_{s=1}^{n-1} A_n B_{ns} + \sum_{t=1}^{n-1} A_t B_{tn}.$$

Therefore identity (2) follows.

Problem 8. Let ϵ be a real parameter and $a_n, b_n \in \mathbb{R}$. Assume that

$$\exp \left(\sum_{n=1}^{\infty} \frac{(i\epsilon)^n b_n}{n!} \right) = \sum_{n=0}^{\infty} \frac{(i\epsilon)^n a_n}{n!} \tag{1}$$

where $a_0 = 1$. Find the relationship between the coefficients a_n and b_n.

Solution 8. An arbitrary term of the exponential function on the left hand side of (1) is given by

$$\frac{1}{k!} \left(\sum_{n=1}^{\infty} \frac{(i\epsilon)^n b_n}{n!} \right)^k = \frac{1}{k!} \left(\sum_{n_1=1}^{\infty} \frac{(i\epsilon)^{n_1} b_{n_1}}{n_1!} \right) \cdots \left(\sum_{n_k=1}^{\infty} \frac{(i\epsilon)^{n_k} b_{n_k}}{n_k!} \right)$$

$$= \frac{1}{k!} \sum_{n_1=1}^{\infty} \sum_{n_2=1}^{\infty} \cdots \sum_{n_k=1}^{\infty} \frac{(i\epsilon)^{n_1+n_2+\cdots+n_k} b_{n_1} b_{n_2} \cdots b_{n_k}}{n_1! n_2! \ldots n_k!}.$$

Therefore

$$\exp\left(\sum_{n=1}^{\infty} \frac{(i\epsilon)^n b_n}{n!}\right) \equiv 1 + \sum_{n=1}^{\infty} \frac{(i\epsilon)^n b_n}{n!} + \frac{1}{2!} \sum_{n_1=1}^{\infty} \sum_{n_2=1}^{\infty} \frac{(i\epsilon)^{n_1+n_2} b_{n_1} b_{n_2}}{n_1! n_2!} + \cdots$$

$$+ \frac{1}{k!} \sum_{n_1=1}^{\infty} \sum_{n_2=1}^{\infty} \cdots \sum_{n_k=1}^{\infty} \frac{(i\epsilon)^{n_1+n_2+\cdots+n_k} b_{n_1} b_{n_2} \cdots b_{n_k}}{n_1! n_2! \cdots n_k!} + \cdots$$

$$= \sum_{n=0}^{\infty} \frac{(i\epsilon)^n a_n}{n!}.$$

Equating terms of the same power in $i\epsilon$ we obtain for the first three terms

$$
\begin{aligned}
(i\epsilon)^1 : & \qquad a_1 = b_1 \\
(i\epsilon)^2 : & \qquad a_2 = b_2 + b_1^2 \\
(i\epsilon)^3 : & \qquad a_3 = b_3 + 3b_2 b_1 + b_1^3.
\end{aligned}
$$

It follows that

$$b_1 = a_1, \qquad b_2 = a_2 - a_1^2, \qquad b_3 = a_3 - 3a_2 a_1 + 2a_1^3.$$

Let X be a random variable with probability density function $f_X(x)$. The n-th moment of X is defined as

$$\langle X^n \rangle := \int x^n f_X(x) dx$$

where the integral is over the entire range of X and $n = 0, 1, 2, \ldots$. Therefore $\langle X^0 \rangle = 1$. The *characteristic function* $\phi_X(k)$ is defined as

$$\phi_X(k) := \sum_{n=0}^{\infty} \frac{(ik)^n \langle X^n \rangle}{n!}.$$

The *cumulant expansion* is given by

$$\phi_X(k) = \exp\left(\sum_{n=1}^{\infty} \frac{(ik)^n C_n(X)}{n!}\right).$$

Thus the equations give the connection between $C_n(X)$ and $\langle X^n \rangle$. The cumulant expansion also plays an important rôle in high temperature expansion in statistical physics.

Problem 9. Show that

$$(\cos\theta + i\sin\theta)^n \equiv \cos(n\theta) + i\sin(n\theta) \qquad (1)$$

where $\theta \in \mathbb{R}$ and $n \in \mathbb{N}$. Identity (1) is sometimes called *De Moivre's theorem*.

Solution 9. We apply the *principle of mathematical induction*. The identity (1) is clearly true for $n = 1$. Assume that the identity is true for k, i.e.

$$(\cos\theta + i\sin\theta)^k \equiv \cos(k\theta) + i\sin(k\theta).$$

Multiplying both sides by $\cos\theta + i\sin\theta$ yields

$$(\cos\theta + i\sin\theta)^{k+1} = (\cos k\theta + i\sin k\theta)(\cos\theta + i\sin\theta)$$

or

$$(\cos\theta + i\sin\theta)^{k+1} = \cos((k+1)\theta) + i\sin((k+1)\theta).$$

Consequently, if the result is true for $n = k$ then it is also true for $n = k+1$. Since the result is true for $n = 1$, it has to be true for $n = 1+1 = 2$, $n = 2+1$ etc. and so has to be true for all $n \in \mathbb{N}$. We have

$$(e^{i\theta})^n \equiv e^{in\theta}$$

where $\theta \in \mathbb{R}$ and $n \in \mathbb{N}$.

Problem 10. Show that every real number $r > 0$ can be represented as the *Cantor series*

$$r = \sum_{\nu=1}^{\infty} \frac{c_\nu}{\nu!} = c_1 + \frac{c_2}{2!} + \frac{c_3}{3!} + \cdots \qquad (1)$$

where

$$0 \le c_\nu \le \nu - 1, \qquad \nu > 1, \qquad c_\nu \in \mathbb{N} \cup \{0\}.$$

Solution 10. We write the real number r in the form

$$r = [r] + \frac{\rho_2}{2} = c_1 + \frac{\rho_2}{2} \qquad (2)$$

where $[r]$ denotes the largest integer $\le r$. Thus $[r] = c_1$ and $\rho_2 < 2$. We set

$$\rho_2 = [\rho_2] + \frac{\rho_3}{3} = c_2 + \frac{\rho_3}{3}, \qquad \rho_{n-1} = [\rho_{n-1}] + \frac{\rho_n}{n} = c_{n-1} + \frac{\rho_n}{n} \qquad (3)$$

and

$$\rho_n = [\rho_n] + \frac{\rho_{n+1}}{n+1} = c_n + \frac{\rho_{n+1}}{n+1}. \qquad (4)$$

Thus we have

$$c_m \leq \rho_m < m \tag{5}$$

for $m = 2, 3, 4, \ldots, n$. Inserting (3) and (4) into (2) yields

$$r = c_1 + \frac{c_2}{2} + \frac{c_3}{2 \cdot 3} + \cdots + \frac{c_n}{n!} + \frac{\rho_{n+1}}{(n+1)!}.$$

Using (5) we find

$$0 \leq r - \left(c_1 + \frac{c_2}{2} + \frac{c_3}{3!} + \cdots + \frac{c_n}{n!} \right) < \frac{1}{n!}.$$

Thus (1) follows.

Problem 11.   Let $z = \exp(i\alpha)$, where $\alpha \in \mathbb{R}$. Let n be a positive integer. Calculate

$$\sum_{k=-n}^{n} z^k.$$

Solution 11.   For $\alpha \neq 0$ we have

$$\sum_{k=-n}^{n} z^k = z^{-n} \sum_{k=0}^{2n} z^k$$

$$= z^{-n} \frac{z^{2n+1} - 1}{z - 1}$$

$$= \frac{z^{n+1} - z^{-n}}{z - 1}$$

$$= \frac{z^{n+1/2} - z^{-(n+1/2)}}{z^{1/2} - z^{-1/2}}$$

$$= \frac{\sin((n+1/2)\alpha)}{\sin(\alpha/2)}.$$

For $\alpha = 0$ we have $2n + 1$.

Problem 12.   Calculate the infinite series

$$\sum_{k=1}^{\infty} \frac{1}{2k^2 - k}.$$

Hint. Use partial fraction decomposition and that

$$\ln(1+x) = \sum_{k=1}^{\infty} (-1)^{k+1} \frac{x^k}{k}, \qquad -1 < x \leq 1.$$

Solution 12. The *partial fraction decomposition* provides

$$\frac{1}{2k^2 - k} \equiv \frac{2}{2k - 1} - \frac{1}{k} \equiv \frac{2}{2k - 1} - \frac{2}{2k}.$$

Therefore

$$\sum_{k=1}^{\infty} \frac{1}{2k^2 - k} = 2 \sum_{k=1}^{\infty} \left(\frac{1}{2k - 1} - \frac{1}{2k} \right) = 2 \left(1 - \frac{1}{2} + \frac{1}{3} - \frac{1}{4} + \cdots \right).$$

We have an alternating series which is convergent. Using the expansion for $\ln(1 + x)$ we find

$$\sum_{k=1}^{\infty} \frac{1}{2k^2 - k} = 2 \ln 2.$$

Problem 13. A series $\sum_{j=0}^{\infty} a_j$, of real or complex numbers, is said to have a $(C, 1)$ sum (a *Cesáro sum*) , s, if $t_n \to s$ as $n \to \infty$, where

$$t_n := \frac{s_0 + s_1 + \cdots + s_n}{n + 1}$$

and the s_k are the partial sums

$$s_k := a_0 + a_1 + \cdots + a_k.$$

Consider the alternating series $\sum_{j=0}^{\infty} (-1)^j = 1 - 1 + 1 - 1 + \cdots$. The series is not convergent. Calculate the Cesáro sum.

Solution 13. We have

$$t_{2n} = \frac{(n + 1)}{2n + 1}, \qquad t_{2n+1} = \frac{(n + 1)}{2n + 2}.$$

Thus $t_{2n} \to 1/2$, $t_{2n+1} \to 1/2$ as $n \to \infty$.

Problem 14. Show that

$$\sum_{k=1}^{\infty} \frac{k}{3 \cdot 5 \cdot 7 \cdots (2k + 1)} = \frac{1}{3} + \frac{2}{3 \cdot 5} + \frac{3}{3 \cdot 5 \cdot 7} + \cdots = \frac{1}{2}. \qquad (1)$$

Solution 14. We consider the partial sums $(n \geq 1)$

$$S_n = \sum_{k=1}^{n} \frac{k}{3 \cdot 5 \cdot 7 \cdots (2k + 1)}$$

and the difference $1/2 - S_n$. We find

$$S_1 = \frac{1}{3} = \frac{1}{2}\left(1 - \frac{1}{3}\right)$$

$$S_2 = \frac{7}{3 \cdot 5} = \frac{1}{2}\left(1 - \frac{1}{3 \cdot 5}\right)$$

$$S_3 = \frac{52}{3 \cdot 5 \cdot 7} = \frac{1}{2}\left(1 - \frac{1}{3 \cdot 5 \cdot 7}\right)$$

suggesting the formula

$$S_n = \frac{1}{2}\left(1 - \frac{1}{3 \cdot 5 \cdots (2n+1)}\right)$$

which can be proved by induction. Thus (1) follows.

Problem 15.   Let $x \in \mathbb{R}$ and $|x| < 1$. Then we know that

$$\sum_{k=0}^{\infty} x^k = \frac{1}{1-x}.$$

(i) Using this result calculate

$$\sum_{k=1}^{\infty} \tanh^{2k}(r), \qquad r \in \mathbb{R}.$$

(ii) Using the result from (i) calculate

$$\sum_{k=1}^{\infty} k \tanh^{2k}(r).$$

Solution 15.   (i) Setting $x = \tanh^2(r)$ we have

$$\sum_{k=1}^{\infty} \tanh^{2k}(r) = \sum_{k=0}^{\infty} \tanh^{2k}(r) - 1$$

$$= \sum_{k=0}^{\infty} x^k - 1$$

$$= \frac{1}{1-x} - 1$$

$$= \frac{x}{1-x}$$

$$= \frac{\tanh^2(r)}{1 - \tanh^2(r)}$$

$$= \sinh^2(r).$$

(ii) Since

$$\frac{d}{dr}\sinh^2(r) = 2\sinh(r)\cosh(r)$$

differentiation of $\tanh^{2k}(r)$ with respect to r provides

$$\frac{d}{dr}\sum_{k=1}^{\infty}\tanh^{2k}(r) = \frac{2}{\cosh^2(r)}\sum_{k=1}^{\infty}k\tanh^{2k-1}(r)$$

$$= \frac{2}{\cosh^2(r)\tanh(r)}\sum_{k=1}^{\infty}k\tanh^{2k}(r)$$

$$= 2\sinh(r)\cosh(r)\,.$$

Thus using $\tanh(r) = \sinh(r)/\cosh(r)$ we finally arrive at

$$\sum_{k=1}^{\infty}k\tanh^{2k}(r) = \sinh^2(r)\cosh^2(r)\,.$$

Problem 16. Let $n \in \mathbb{N}$ and $x \in \mathbb{R}$ with $|x| < 1$. Show that $(1+x)^n$ can be expressed using the exponential function and a series expansion for $\ln(1+x)$.

Solution 16. We have

$$(1+x)^n = \exp(n\ln(1+x)) = \exp\left(n\sum_{k=1}^{\infty}\frac{(-1)^{k-1}x^k}{k}\right)\,.$$

Problem 17. Find the sum of the series

$$\sum_{j=0}^{\infty}\frac{1}{(4j)!}\,.$$

Hint. Start with the expansions for $\cos(x)$ and $\cosh(x)$.

Solution 17. We have

$$\cos(x) = \sum_{j=0}^{\infty}\frac{(-1)^n x^{2j}}{(2j)!}, \qquad \cosh(x) = \sum_{j=0}^{\infty}\frac{x^{2j}}{(2j)!}\,.$$

Thus

$$\cos(x) + \cosh(x) = 2\sum_{j=0}^{\infty}\frac{x^{4j}}{(4j)!}\,.$$

If follows that

$$\sum_{j=0}^{\infty} \frac{1}{(4j)!} = \frac{1}{2}(\cos(1) + \cosh(1)).$$

Problem 18.   Let $\phi \in \mathbb{R}$. Calculate the infinite series

$$f(\alpha) = \sum_{j=1}^{\infty} \sin\left(\frac{2\phi}{3^j}\right) \sin\left(\frac{\phi}{3^j}\right).$$

Hint. Use the identity

$$\sin(\phi_1)\sin(\phi_2) \equiv \frac{1}{2}\cos(\phi_1 - \phi_2) - \frac{1}{2}\cos(\phi_1 + \phi_2).$$

Solution 18.   Using the identity we have for the kth partial sum

$$S_k = \sum_{j=1}^{k} \sin\left(\frac{2\phi}{3^j}\right) \sin\left(\frac{\phi}{3^j}\right)$$

$$= \sum_{j=1}^{k} \left(\frac{1}{2}\cos\left(\frac{\phi}{3^j}\right) - \frac{1}{2}\cos\left(\frac{\phi}{3^{j-1}}\right)\right)$$

$$= \frac{1}{2}\cos\left(\frac{\phi}{3^k}\right) - \frac{1}{2}\cos(\phi).$$

Thus

$$\sum_{j=1}^{\infty} \sin\left(\frac{2\phi}{3^j}\right) \sin\left(\frac{\phi}{3^j}\right) = \lim_{k\to\infty}\left(\frac{1}{2}\cos\left(\frac{\phi}{3^k}\right) - \frac{1}{2}\cos\phi\right) = \frac{1}{2}(1 - \cos\phi).$$

Problem 19.   Let N be a positive integer. Consider the set

$$S := \left\{ \frac{2\pi k}{2N} : k = -N, -N+1, \ldots, 0, \ldots, N-1 \right\}.$$

Calculate the sum

$$\sum_{k \in S} \cos(k).$$

Solution 19.   Since $\cos(-x) = \cos(x)$ and $\cos(x + \pi) = -\cos(x)$ we find

$$\sum_{k \in S} \cos(k) = 0.$$

Problem 20. Let $z = e^{i\pi/n}$ with $n \geq 2$. Calculate the sum

$$S_n(z) = 2\sum_{k=1}^{n} kz^{2k-1} .$$

Solution 20. We set

$$R_n(z) = \sum_{k=1}^{n} z^{2k}$$

and $z \neq 1$. From $R_n(z) - z^2 R_n(z)$ we obtain

$$R_n(z) = \frac{z^2 - z^{2n+2}}{1 - z^2} .$$

Differentiation of $R_n(z)$ with respect to z yields

$$\frac{dR_n(z)}{dz} = \frac{2z - (2n+2)z^{2n+1} + 2nz^{2n+3}}{(1 - z^2)^2} = 2\sum_{k=1}^{n} kz^{2k-1} = S_n(z) .$$

Since $z^{2n} = 1$ we arrive at

$$S_n(z) = -\frac{2nz}{1 - z^2} = -\frac{2n}{1/z - z} = \frac{2n}{e^{i\pi/n} - e^{-i\pi/n}} = \frac{n}{i\sin(\pi/n)} .$$

Problem 21. Let a, b be nonnegative integers, $b > a$ and $x \in \mathbb{R}$. Calculate

$$\sum_{j=a}^{b} e^{jx} .$$

Solution 21. For $x \neq 0$ we have

$$\sum_{j=a}^{b} e^{jx} = e^{ax}(1 + e^x + e^{2x} + \cdots + e^{(b-a)x})$$

$$= e^{ax}\frac{1 - e^{(b-a+1)x}}{1 - e^x}$$

$$-\frac{e^{ax}}{1 - e^x} + \frac{e^{bx}}{1 - e^{-x}} .$$

For $x = 0$ we have $b - a + 1$.

Problem 22.   Let f_k ($k = 1, 2, \ldots, n$) be differentiable functions. Assume that $f_k \neq 0$ for $k = 1, 2, \ldots, n$. Let

$$S(x) = \prod_{k=1}^{n} f_k(x) \equiv f_1(x)f_2(x) \cdots f_n(x).\qquad(1)$$

Calculate the derivative of S with respect to x.

Solution 22.   We apply the *product rule* for differentiation

$$\frac{d}{dx}(h(x)g(x)) = \frac{dh}{dx}g + h\frac{dg}{dx}$$

to S, i.e.

$$\frac{dS}{dx} = \frac{d}{dx}\prod_{k=1}^{n} f_k(x).$$

Thus

$$\frac{dS}{dx} \equiv \frac{df_1}{dx}f_2 \cdots f_n + f_1\frac{df_2}{dx} \cdots f_n + \cdots + f_1 f_2 \cdots \frac{df_n}{dx}$$

or

$$\frac{dS}{dx} \equiv \frac{1}{f_1}\frac{df_1}{dx}S(x) + \frac{1}{f_2}\frac{df_2}{dx}S(x) + \cdots + \frac{1}{f_n}\frac{df_n}{dx}S(x).$$

Consequently,

$$\frac{dS}{dx} = S(x)\sum_{j=1}^{n}\frac{1}{f_j}\frac{df_j}{dx} \equiv S(x)\sum_{j=1}^{n}\frac{d}{dx}(\ln f_j(x)).$$

Problem 23.   Let n be a positive integer with $n \geq 2$ and $\lambda := \exp(2\pi i/n)$. We have

$$\prod_{j=1}^{n-1}(x - \lambda^j) = \frac{x^n - 1}{x - 1}.\qquad(1)$$

Show that

$$\sum_{j=1}^{n-1}\frac{1}{x - \lambda^j} = \frac{d}{dx}\ln\left(\frac{x^n - 1}{x - 1}\right).\qquad(2)$$

Solution 23.   Taking the logarithm of (1) we obtain

$$\sum_{j=1}^{n-1}\ln(x - \lambda^j) = \ln\left(\frac{x^n - 1}{x - 1}\right).\qquad(3)$$

Taking the derivative of (3) with respect to x yields (2).

Chapter 3

Discrete Fourier Transform

Let N be a positive integer, $N \geq 2$. The discrete one-dimensional Fourier transform for a given sequence

$$x(n), \qquad n = 0, 1, 2, \ldots, N-1$$

is defined by

$$\hat{x}(k) := \frac{1}{N} \sum_{n=0}^{N-1} x(n) e^{-i2\pi kn/N}$$

where $k = 0, 1, 2, \ldots, N-1$. The discrete Fourier transform is linear.

Problem 1. (i) Calculate

$$\sum_{k=0}^{N-1} e^{i2\pi k(n-m)/N} \qquad (1)$$

where $n, m \in \mathbb{N} \cup \{0\}$, $N \in \mathbb{N}$ and $0 \leq n, m \leq N-1$.
(ii) Find the discrete inverse Fourier transform.

Solution 1. (i) If $n = m$, then

$$\sum_{k=0}^{N-1} 1 = N.$$

23

If $n \neq m$, then $n - m = q$, where $0 < q \leq N - 1$. Then the sum (1) is a geometric series. Therefore

$$\sum_{k=0}^{N-1} e^{i2\pi kq/N} = 0.$$

Consequently,

$$\sum_{k=0}^{N-1} e^{i2\pi k(n-m)/N} = N\delta_{nm}$$

where δ_{nm} denotes the *Kronecker delta*.

(ii) From the definition of the discrete Fourier transform it follows that

$$\hat{x}(k)e^{i2\pi km/N} = \frac{1}{N}\sum_{n=0}^{N-1} x(n)e^{i2\pi k(m-n)/N}.$$

Summation over k of both sides gives

$$\sum_{k=0}^{N-1}\hat{x}(k)e^{i2\pi km/N} = \frac{1}{N}\sum_{k=0}^{N-1}\sum_{n=0}^{N-1} x(n)e^{i2\pi k(m-n)/N}$$

$$= \frac{1}{N}\sum_{n=0}^{N-1} x(n)\sum_{k=0}^{N-1} e^{i2\pi k(m-n)/N}.$$

Using the result from (i) we obtain

$$\sum_{k=0}^{N-1}\hat{x}(k)e^{i2\pi km/N} = \frac{1}{N}\sum_{n=0}^{N-1} x(n)N\delta_{mn} = x(m).$$

Therefore

$$x(n) = \sum_{k=0}^{N-1}\hat{x}(k)e^{i2\pi kn/N}.$$

Problem 2.   Let
$$x(n) = \cos(2\pi n/N) \tag{1}$$
where $N = 8$ and $n = 0, 1, 2, \ldots, N - 1$. Find $\hat{x}(k)$ ($k = 0, 1, 2, \ldots, N - 1$), i.e., find the discrete Fourier transform.

Solution 2.   We have

$$\hat{x}(k) = \frac{1}{8}\sum_{n=0}^{7}\cos\left(\frac{2\pi n}{8}\right)e^{-i2\pi kn/8}. \tag{2}$$

Using the identity $\cos(2\pi n/8) \equiv (e^{i2\pi n/8} + e^{-i2\pi n/8})/2$ we have

$$\hat{x}(k) = \frac{1}{16} \sum_{n=0}^{7} (e^{i2\pi n(1-k)/8} + e^{-i2\pi n(1+k)/8}).$$

Consequently,

$$\hat{x}(k) = \begin{cases} \dfrac{1}{2} & \text{for} \quad k = 1 \\ \dfrac{1}{2} & \text{for} \quad k = 7 \\ 0 & \text{otherwise.} \end{cases}$$

Problem 3. Consider the sequence

$$\mathbf{x} = (x(0), x(1), \ldots, x(N-1))$$

and the sequence of the discrete Fourier transform

$$\hat{\mathbf{x}} = (\hat{x}(0), \hat{x}(1), \ldots, \hat{x}(N-1)).$$

We define the scalar products

$$(\mathbf{x}, \mathbf{x}) := \sum_{n=0}^{N-1} \bar{x}(n)x(n), \qquad (\hat{\mathbf{x}}, \hat{\mathbf{x}}) := \sum_{k=0}^{N-1} \bar{\hat{x}}(k)\hat{x}(k) \tag{1}$$

where $\bar{x}(n)$ denotes the complex conjugate of $x(n)$. Show that

$$(\mathbf{x}, \mathbf{x}) = N(\hat{\mathbf{x}}, \hat{\mathbf{x}}).$$

Solution 3. From (1) we have

$$(\mathbf{x}, \mathbf{x}) = \sum_{n=0}^{N-1} \bar{x}(n)x(n) = \sum_{k=0}^{N-1} \sum_{l=0}^{N-1} \bar{\hat{x}}(k)\hat{x}(l) \sum_{n=0}^{N-1} e^{i2\pi n(l-k)/N}.$$

Using

$$\sum_{n=0}^{N-1} e^{i2\pi n(l-k)/N} = N\delta_{lk}$$

we find

$$(\mathbf{x}, \mathbf{x}) = N \sum_{k=0}^{N-1} \bar{\hat{x}}(k)\hat{x}(k).$$

Thus $(\mathbf{x}, \mathbf{x}) = N(\hat{\mathbf{x}}, \hat{\mathbf{x}})$.

Problem 4.   Let $x(m)$, $y(m)$ be two sequences with $m = 0, 1, \ldots, N - 1$. Show that the discrete Fourier transform of a circular convolution

$$(x \odot y)(n) = \sum_{m=0}^{N-1} x(m)y(n-m) = \sum_{m=0}^{N-1} x(n-m)y(m)$$

is the product of the two discrete Fourier transform of $x(n)$ and $y(m)$. For a circular convolution we have

$$x(m + N) = x(m), \qquad y(m + N) = y(m).$$

Solution 4.   Let $u(n) = (x \odot y)(n)$. Then

$$\hat{u}(k) = \frac{1}{N} \sum_{n=0}^{N-1} u(n)e^{-i2\pi kn/N}$$

$$= \frac{1}{N} \sum_{n=0}^{N-1} \sum_{m=0}^{N-1} x(m)y(n-m)e^{-i2\pi kn/N}.$$

Using $n - m = p(\operatorname{mod} N)$ we obtain

$$\hat{u}(k) = \frac{1}{N} \sum_{m=0}^{N-1} \sum_{\substack{p+m=0 \ \operatorname{mod} N}}^{N-1} x(m)y(p)e^{-i2\pi k(p+m)/N}$$

$$= \left(\frac{1}{N} \sum_{m=0}^{N-1} x(m)e^{-i2\pi km/N} \right) \sum_{\substack{p+m=0 \ \operatorname{mod} N}}^{N-1} y(p)e^{-i2\pi kp/N}$$

$$= \hat{x}(k) \sum_{\substack{p+m=0 \ \operatorname{mod} N}}^{N-1} y(p)e^{-i2\pi kp/N}.$$

Next we have to calculate the sum

$$\sum_{\substack{p+m=0 \ \operatorname{mod} N}}^{N-1} y(p)e^{-i2\pi kp/N}$$

Note that we have mod N summation. This means we sum over all terms $p + m = j \operatorname{mod} N$. For example if $N = 4$ we sum over all combinations (p, m) with $p < N$, $m < N$, i.e.

$$\begin{array}{cccc}
(0,0), & (1,3), & (2,2), & (3,1) \\
(0,1), & (1,0), & (2,3), & (3,2) \\
(0,2), & (2,0), & (1,1), & (3,3) \\
(0,3), & (3,0), & (1,2), & (2,1)
\end{array}$$

Thus

$$\sum_{p+m=0 \bmod N}^{N-1} y(p)e^{-i2\pi kp/N} = N \sum_{p=0}^{N-1} y(p)e^{-i2\pi kp/N} = N^2 \hat{y}(k).$$

Thus the result follows.

Problem 5.   The *two-dimensional discrete Fourier transform* of a two-dimensional array is computed by first performing a one-dimensional Fourier transform of the rows of the array and then a one-dimensional Fourier transform of the columns of the result (or vice versa). The absolute value of the Fourier coefficients does not change under a translation of the two-dimensional pattern. The Fourier transform also has the property that it preserves angles, that is, similar patterns in the original domain are also similar in the Fourier domain.

Definition.   Let $x(n_1, n_2)$ denote an array of real values, where n_1, n_2 are integers such that $0 \le n_1 \le N_1 - 1$ and $0 \le n_2 \le N_2 - 1$. The two-dimensional discrete Fourier transform $\hat{x}(k_1, k_2)$ of $x(n_1, n_2)$ is defined by

$$\hat{x}(k_1, k_2) := \frac{1}{N_1} \frac{1}{N_2} \sum_{n_1=0}^{N_1-1} \sum_{n_2=0}^{N_2-1} x(n_1, n_2) \exp\left(-\frac{2\pi}{N_1} i n_1 k_1 - \frac{2\pi}{N_2} i n_2 k_2\right)$$

where $0 \le k_1 \le N_1 - 1$ and $0 \le k_2 \le N_2 - 1$.

Consider two arrays $x(n_1, n_2)$ and $y(n_1, n_2)$ of real values. Assume that

$$y(n_1, n_2) = x(n_1 + d_1, n_2 + d_2)$$

where d_1 and d_2 are two given integers. The addition $n_1 + d_1$ is performed modulo N_1 and $n_2 + d_2$ is performed modulo N_2 (torus). Show that

$$\|\hat{y}(k_1, k_2)\| = \|\hat{x}(k_1, k_2)\|.$$

Solution 5.   We have

$$\hat{y}(k_1, k_2) = \frac{1}{N_1} \frac{1}{N_2} \sum_{n_1=0}^{N_1-1} \sum_{n_2=0}^{N_2-1} x(n_1+d_1, n_2+d_2) \exp\left(-\frac{2\pi}{N_1} i n_1 k_1 - \frac{2\pi}{N_2} i n_2 k_2\right).$$

With the change of indices $n'_1 = (n_1 + d_1) \bmod N_1$, $n'_2 = (n_2 + d_2) \bmod N_2$ we obtain

$$\hat{y}(k_1, k_2) = \frac{1}{N_1} \frac{1}{N_2} \sum_{n'_1=0}^{N_1-1} \sum_{n'_2=0}^{N_2-1} x(n'_1, n'_2) \exp\left(2\pi i \left(-\sum_{j=1}^{2} \frac{n'_j k_j}{N_j} + \sum_{j=1}^{2} \frac{d_j k_j}{N_j}\right)\right).$$

This can be written as

$$\hat{y}(k_1, k_2) = \exp\left(\frac{2\pi}{N_1} i d_1 k_1\right) \exp\left(\frac{2\pi}{N_2} i d_2 k_2\right) \hat{x}(k_1, k_2).$$

This expression tells us that the Fourier coefficients of the array $y(n_1, n_2)$ are the same as the coefficients of the array $x(n_1, n_2)$ except for a phase factor

$$\exp\left(-\frac{2\pi}{N} i d_1 k_1\right) \exp\left(-\frac{2\pi}{N} i d_2 k_2\right).$$

Taking the absolute value results in

$$\|\hat{y}(k_1, k_2)\| = \|\hat{x}(k_1, k_2)\|.$$

Thus the absolute values of the Fourier coefficients for both patterns are identical.

Problem 6.   The discrete one-dimensional *cosine transform* of the given sequence $x(n)$, $n = 0, 1, 2, \ldots, N - 1$ is defined by

$$C(0) := \frac{1}{\sqrt{N}} \sum_{n=0}^{N-1} x(n)$$

$$C(k) := \sqrt{\frac{2}{N}} \sum_{n=0}^{N-1} x(n) \cos\left(\frac{(2n+1)k\pi}{2N}\right), \qquad k = 1, 2, \ldots, N - 1.$$

Find the inverse discrete one-dimensional cosine transform.

Solution 6.   The inverse one-dimensional cosine transform is given by

$$x(n) = \frac{1}{\sqrt{N}} C(0) + \sqrt{\frac{2}{N}} \sum_{k=1}^{N-1} C(k) \cos\left(\frac{(2n+1)k\pi}{2N}\right).$$

To find the inverse cosine transform we apply the identity

$$\cos(\alpha)\cos(\beta) \equiv \frac{1}{2}\cos(\alpha + \beta) + \frac{1}{2}\cos(\alpha - \beta)$$

with the special case $\cos^2(\alpha) \equiv (\cos(2\alpha) + 1)/2$. Furthermore we use that

$$\sum_{n=0}^{N-1} \cos\left(\frac{(2n+1)k\pi}{2N}\right) = 0, \qquad k = 1, 2, \ldots, N - 1$$

and

$$\frac{1}{\sqrt{N}} \sum_{n=0}^{N-1} C(0) = \sqrt{N} C(0).$$

Chapter 4

Algebraic and Transcendental Equations

A complex polynomial is a complex function of the form

$$p(z) = c_n z^n + c_{n-1} z^{n-1} + \cdots + c_0$$

where $c_0, c_1, \ldots, c_n$ are complex numbers and n is a natural number. A root, or zero, of this polynomial is a complex number w such that $p(w) = 0$. The fundamental theorem of algebra states that any complex polynomial must have a complex root.

Problem 1. Let $n \in \mathbb{N}$. Solve

$$z^n = 1. \tag{1}$$

Solution 1. Since every complex number z can be written as $z = re^{i\phi}$ we have

$$z^n = r^n e^{in\phi}.$$

Taking (1) into account we find $r = 1$ and

$$e^{in\phi} \equiv \cos(n\phi) + i\sin(n\phi) = 1.$$

Therefore

$$z_k = \cos\left(\frac{2\pi k}{n}\right) + i\sin\left(\frac{2\pi k}{n}\right)$$

are the roots of 1, where $k = 0, 1, \ldots, n - 1$.

Problem 2.   Find all solutions of

$$x^4 + x^3 + x^2 + x + 1 = 0. \tag{1}$$

Solution 2.   **Method 1.** Equation (1) can be solved by dividing by x^2, substituting $y = x + 1/x$, and then applying the quadratic formula. Thus, we have

$$x^2 + \frac{1}{x^2} + x + \frac{1}{x} + 1 = 0$$

$$\left(x^2 + 2 + \frac{1}{x^2}\right) + \left(x + \frac{1}{x}\right) + (1 - 2) = 0$$

$$\left(x + \frac{1}{x}\right)^2 + \left(x + \frac{1}{x}\right) - 1 = 0$$

$$y^2 + y - 1 = 0.$$

The roots of this equation are

$$y_1 = \frac{-1 + \sqrt{5}}{2}, \qquad y_2 = \frac{-1 - \sqrt{5}}{2}.$$

It remains to determine x by solving the two equations

$$x + \frac{1}{x} = y_1, \quad \text{and} \quad x + \frac{1}{x} = y_2$$

which are equivalent to

$$x^2 - y_1 x + 1 = 0, \quad \text{and} \quad x^2 - y_2 x + 1 = 0.$$

The four roots found by solving these are

$$x_1 = \frac{-1 + \sqrt{5}}{4} + i\frac{\sqrt{10 + 2\sqrt{5}}}{4}, \quad x_2 = \frac{-1 + \sqrt{5}}{4} - i\frac{\sqrt{10 + 2\sqrt{5}}}{4}$$

$$x_3 = \frac{-1 - \sqrt{5}}{4} + i\frac{\sqrt{10 - 2\sqrt{5}}}{4}, \quad x_4 = \frac{-1 - \sqrt{5}}{4} - i\frac{\sqrt{10 - 2\sqrt{5}}}{4}.$$

Method 2. Another approach to this problem is to multiply each side of the original equation by $x - 1$. Since

$$(x - 1)(x^4 + x^3 + x^2 + x + 1) = x^5 - 1$$

an equivalent problem is to find all x (other than $x = 1$) which satisfy $x^5 = 1$. These are the five fifth roots of unity, given by

$$x_1 = \cos\frac{2}{5}\pi + i\sin\frac{2}{5}\pi, \quad x_2 = \cos\frac{4}{5}\pi + i\sin\frac{4}{5}\pi$$

$$x_3 = \cos\frac{6}{5}\pi + i\sin\frac{6}{5}\pi, \quad x_4 = \cos\frac{8}{5}\pi + i\sin\frac{8}{5}\pi$$

and $x_5 = 1$.

Problem 3. Given a set of N real numbers $x_1, x_2, \ldots, x_N$. It is often useful to express the sum of the j powers

$$s_j = x_1^j + x_2^j + \cdots + x_N^j, \qquad j = 0, 1, 2, \ldots$$

in terms of the *elementary symmetric functions*

$$\sigma_1 = \sum_{i=1}^{N} x_i$$

$$\sigma_2 = \sum_{i<j}^{N} x_i x_j$$

$$\sigma_3 = \sum_{i<j<k}^{N} x_i x_j x_k$$

$$\vdots$$

$$\sigma_N = x_1 x_2 \cdots x_N.$$

Consider the special case with three numbers x_1, x_2, x_3. Then the elementary symmetric functions are given by

$$\sigma_1 = x_1 + x_2 + x_3, \quad \sigma_2 = x_1 x_2 + x_1 x_3 + x_2 x_3, \quad \sigma_3 = x_1 x_2 x_3.$$

We know that the elementary symmetric functions are the coefficients (up to sign) of the polynomial with the roots x_1, x_2, x_3. In other words the values of x_1, x_2, x_3 each satisfy the polynomial equation

$$x^3 - \sigma_1 x^2 + \sigma_2 x - \sigma_3 = 0. \tag{1}$$

Find a recursion relation for

$$s_j := x_1^j + x_2^j + x_3^j, \qquad j = 0, 1, 2, \ldots$$

and give the initial values. Calculate s_3 and s_4.

Solution 3.   We can multiply equation (1) by any power of x to provide the equation

$$x^j - \sigma_1 x^{j-1} + \sigma_2 x^{j-2} - \sigma_3 x^{j-3} = 0, \quad j = 3, 4, \ldots .$$

Thus we find the recursion relation

$$s_j - \sigma_1 s_{j-1} + \sigma_2 s_{j-2} - \sigma_3 s_{j-3} = 0, \quad j = 3, 4, \ldots \qquad (2)$$

with the inital conditions

$$s_0 = 3, \quad s_1 = \sigma_1, \quad s_2 = \sigma_1^2 - 2\sigma_2 .$$

Inserting these initial conditions into the recursion relation (2) yields

$$s_3 = \sigma_1^3 - 3\sigma_1\sigma_2 + 3\sigma_3$$

and

$$s_4 = \sigma_1^4 - 4\sigma_1^2\sigma_2 + 2\sigma_2^2 + 4\sigma_1\sigma_3 .$$

Since σ_1 is of degree 1, σ_2 is of degree 2, and σ_3 is of degree 3, each of these expressions is homogeneous.

Problem 4.   Let L be a given positive real number. Solve the system of two coupled nonlinear equations

$$1 = x^2 L + 2x^2 y + Lx^2 y^2, \qquad 0 = -x^2 + 2x^2 y + Lx^2 y^2 .$$

Solution 4.   Subtracting the second equation from the first equation we obtain $1 = x^2(1 + L)$ or $x^2 = 1/(1 + L)$. Inserting $x^2 = 1/(1 + L)$ into the second equation we obtain

$$x = \frac{\pm 1}{\sqrt{L + 1}}, \qquad y = \frac{-1 \pm \sqrt{L + 1}}{L} .$$

Problem 5.   Find the solution of the equation

$$\cos x + \cos(3x) + \cdots + \cos((2n - 1)x) = 0 \qquad (1)$$

where $n \in \mathbb{N}$.

Solution 5.   Obviously, (1) is invariant under $x \rightarrow x + \pi$. Now (1) can be written in the form

$$\Re(e^{ix} + e^{3ix} + \cdots + e^{(2n-1)ix}) = 0$$

where $\Re$ denotes the real part of a complex number. The sum on the left-hand side is a geometric series. Therefore we obtain for the real part

$$\frac{\sin(2nx)}{2\sin x} = 0.$$

Consequently,

$$x = \frac{\pi}{2n}, \quad \frac{2\pi}{2n}, \quad \frac{3\pi}{2n}, \quad \cdots \quad , \frac{(2n-1)\pi}{2n} \quad \text{modulo } \pi.$$

The solution of this problem also gives the extrema of the function

$$f(x) = \sin x + \frac{1}{3}\sin(3x) + \cdots + \frac{1}{2n-1}\sin((2n-1)x).$$

The function f is a *Fourier approximation*.

Problem 6.   (i) Solve the equation

$$\sin\left((2n+1)\phi\right) = 0 \tag{1}$$

where $n \in \mathbb{Z}$ and $\phi \in \mathbb{R}$.
(ii) Solve the equation

$$\sin z = 0 \tag{2}$$

where $z \in \mathbb{C}$.

Solution 6.   (i) From $\sin x = 0$ with $x \in \mathbb{R}$ we obtain $x = k\pi$ with $k \in \mathbb{Z}$. Therefore $(2n+1)\phi = k\pi$ or

$$\phi = \frac{k\pi}{2n+1}.$$

(ii) Since

$$\sin z \equiv \sin(x + iy)$$
$$\equiv \sin x \cos(iy) + \cos x \sin(iy)$$
$$\equiv \sin x \cosh y + i \cos x \sinh y$$

we have

$$\sin x \cosh y + i \cos x \sinh y = 0.$$

It follows that

$$\sin x \cosh y = 0, \qquad \cos x \sinh y = 0.$$

Since $\cosh y \neq 0$ for $y \in \mathbb{R}$, the first equation can only be satisfied if $x = k\pi$, where $k \in \mathbb{Z}$. Since $\cos(k\pi) \neq 0$ if $k \in \mathbb{Z}$, the second equation can only be satisfied if $y = 0$. Consequently, the solution to $\sin(z) = 0$ is $z = x + iy$ with $x = k\pi$ and $y = 0$.

Problem 7.   Let $0 < \theta < 1$ and

$$\sum_{k=1}^{\infty}\sum_{j=0}^{k-1} \theta^{j+k+2} = 1.$$

Find θ.

Solution 7.   We have

$$\sum_{k=1}^{\infty}\sum_{j=0}^{k-1} \theta^{j+k+2} = \sum_{k=1}^{\infty} \theta^{k+2} \sum_{j=0}^{k-1} \theta^j$$

$$= \sum_{k=1}^{\infty} \theta^{k+2} \left(\frac{1 - \theta^k}{1 - \theta}\right)$$

$$= \frac{\theta^2}{1 - \theta}\left(\sum_{k=1}^{\infty}\theta^k - \sum_{k=1}^{\infty}(\theta^2)^k\right)$$

$$= \frac{\theta^2}{1 - \theta}\left(\left(\frac{1}{1-\theta} - 1\right) - \left(\frac{1}{1-\theta^2} - 1\right)\right)$$

$$= \frac{\theta^2}{1 - \theta}\left(\frac{\theta}{1-\theta} - \frac{\theta^2}{1-\theta^2}\right)$$

$$= \frac{\theta^3}{(1 - \theta)(1 - \theta^2)}$$

$$= 1.$$

Thus we obtain the quadratic equation $\theta^2 + \theta - 1 = 0$ with the solution (golden mean number)

$$\theta = \frac{-1 + \sqrt{5}}{2}.$$

Problem 8.   A curve $f(x, y) = 0$ admits a *rational parametrization* if there exist rational functions α and β such

$$f(\alpha(t), \beta(t)) = 0$$

i.e.

$$f(x, y) = 0 \Leftrightarrow \begin{cases} x = \alpha(t) \\ y = \beta(t) \end{cases}$$

Such a curve $f(x, y) = 0$ is called *unicursal*. The curve $x^n + y^n - 1 = 0$ is not unicursal for $n \geq 3$, otherwise *Fermat-Wiles* would be wrong.

(i) Find the rational parametrization for the unit circle $x^2 + y^2 - 1 = 0$.
(ii) Find the rational parametrization for the *nodal cubic* $y^3 + x^3 - xy = 0$.

Solution 8. (i) The circle admits the rational parametrization

$$\alpha(t) = \frac{1 - t^2}{1 + t^2}, \qquad \beta(t) = \frac{2t}{1 + t^2}.$$

With $t = \tan(\theta/2)$ we obtain $\alpha(\theta) = \cos\theta$ and $\beta(\theta) = \sin\theta$.
(ii) The nodal cubic admits the rational parametrization

$$\alpha(t) = \frac{t}{1 + t^3}, \qquad \beta(t) = \frac{t^2}{1 + t^3}.$$

Problem 9. A *diophantine equation* is a polynomial equation with integral coefficients as unknowns. Find the integral solutions of

$$x^2 + y^2 = z^2. \tag{1}$$

Solution 9. Obviously $x = y = z = 0$ is a solution. With $z \neq 0$ we obtain

$$\left(\frac{x}{z}\right)^2 + \left(\frac{y}{z}\right)^2 = 1, \qquad x/z,\ y/z \in \mathbb{Q}.$$

Thus

$$\left(\frac{x}{z}\right)^2 + \left(\frac{y}{z}\right)^2 = 1 \Leftrightarrow \begin{cases} \dfrac{x}{z} = \dfrac{1 - t^2}{1 + t^2} \\[2mm] \dfrac{y}{z} = \dfrac{2t}{1 + t^2} \end{cases}$$

and

$$((x/z), (x/y)) \in \mathbb{Q} \times \mathbb{Q} \Leftrightarrow t = y/(1 + x) \in \mathbb{Q}.$$

Let $t = a/b$ $(a, b \in \mathbb{Z})$ and $b \neq 0$. Then the integral solutions of $x^2 + y^2 = z^2$ are

$$x = a^2 - b^2, \qquad y = 2ab, \qquad z = a^2 + b^2$$

for all $a, b \in \mathbb{Z}$.

Problem 10. Solve for x the congruence

$$9x + 5 \equiv 10 \bmod 11.$$

Solution 10. Subtract 5 from both sides of the congruence

$$9x \equiv 5 \bmod 11.$$

Since $\gcd(9, 11) = 1$, division by 9 is valid. Before we can perform the division, we need to choose a suitable multiple of 11 to add to the right-hand side in order to obtain a multiple of 9. We choose 22. Thus

$$9x \equiv 27 \bmod 11.$$

Now we divide by 9 providing

$$x \equiv 3 \bmod 11.$$

Thus the solution consists of all integers of the form $11n + 3$, where $n \in \mathbb{Z}$.

Problem 11.   A real-valued function f, defined on the rational numbers $\mathbb{Q}$, satisfies

$$f(x + y) = f(x) + f(y)$$

for all rational x and y. Prove that

$$f(x) = f(1) \cdot x$$

for all rational x.

Solution 11.   Let $n \in \mathbb{N}$. Then

$$f(1) = f \underbrace{\left(\frac{1}{n} + \cdots + \frac{1}{n} \right)}_{n\times}.$$

Therefore

$$f\left(\frac{1}{n} \right) = \frac{1}{n} f(1).$$

Any rational number can be represented as m/n with m an integer and $n \in \mathbb{N}$. Then

$$f\left(\frac{m}{n} \right) = f \underbrace{\left(\frac{1}{n} + \cdots + \frac{1}{n} \right)}_{m\times}.$$

It follows that

$$f\left(\frac{m}{n} \right) = \underbrace{f\left(\frac{1}{n} \right) + \cdots + f\left(\frac{1}{n} \right)}_{m\times}.$$

Finally

$$f\left(\frac{m}{n} \right) = mf\left(\frac{1}{n} \right) = \frac{m}{n} f(1).$$

Chapter 5

Vector and Matrix Calculations

Let $\mathcal{F}$ be a field, for example the set of real numbers $\mathbb{R}$ or the set of complex numbers $\mathbb{C}$. Let $m, n \geq 1$ be two integers. An array A of numbers in $\mathcal{F}$

$$
\begin{pmatrix}
a_{11} & a_{12} & a_{13} & \cdots & a_{1n} \\
a_{21} & a_{22} & a_{23} & \cdots & a_{2n} \\
\vdots & \vdots & \vdots & \ddots & \vdots \\
a_{m1} & a_{m2} & a_{m3} & \cdots & a_{mn}
\end{pmatrix} = (a_{ij})
$$

is called an $m \times n$ *matrix* with entry a_{ij} in the ith row and jth column. A *row vector* is a $1 \times n$ matrix. A *column vector* is an $n \times 1$ matrix. If $a_{ij} = 0$ for all i, j, then A is called a zero matrix.

Problem 1. Consider the normalized vectors in $\mathbb{R}^2$ ($\theta_1, \theta_2 \in [0, 2\pi)$)

$$
\begin{pmatrix} \cos\theta_1 \\ \sin\theta_1 \end{pmatrix}, \qquad \begin{pmatrix} \cos\theta_2 \\ \sin\theta_2 \end{pmatrix}.
$$

Find the condition on θ_1 and θ_2 such that the vector in $\mathbb{R}^2$

$$
\begin{pmatrix} \cos\theta_1 \\ \sin\theta_1 \end{pmatrix} + \begin{pmatrix} \cos\theta_2 \\ \sin\theta_2 \end{pmatrix}
$$

is normalized.

Solution 1.   From the condition that the vector

$$\begin{pmatrix} \cos\theta_1 + \cos\theta_2 \\ \sin\theta_1 + \sin\theta_2 \end{pmatrix}$$

is normalized it follows that

$$(\sin\theta_1 + \sin\theta_2)^2 + (\cos\theta_1 + \cos\theta_2)^2 = 1.$$

Thus we have

$$\sin\theta_1\sin\theta_2 + \cos\theta_1\cos\theta_2 = -\frac{1}{2}.$$

It follows that

$$\cos(\theta_1 - \theta_2) = -\frac{1}{2}.$$

Therefore, $\theta_1 - \theta_2 = 2\pi/3$ or $\theta_1 - \theta_2 = 4\pi/3$.

Problem 2.   Can one find five vectors $\mathbf{v}_j$ ($j = 1, 2, 3, 4, 5$) in $\mathbb{R}^3$ such that the angle between any two of these vectors is obtuse? An *obtuse angle* θ is one that is larger then $\pi/2$ but less than π. We can set $\mathbf{v}_5 = (0\ 0\ -1)^T$ without loss of generality.

Solution 2.   We set $\mathbf{v}_j = (v_{j1}\ v_{j2}\ v_{j3})^T$. Suppose that all angles are obtuse. Then we have for the scalar product

$$\mathbf{v}_j^T\mathbf{v}_k = |\mathbf{v}_j||\mathbf{v}_k|\cos(\theta_{jk})$$

where θ_{jk} is the (obtuse) angle between $\mathbf{v}_j$ and $\mathbf{v}_k$. Taking $k = 5$ we have $v_{j3} > 0$ for $j = 1, 2, 3, 4$. Consider now the projection $\mathbf{v}_j \to \mathbf{w}_j = (v_{j1}\ v_{j2}\ 0)^T$. Since the $\mathbf{w}_j$ span at most a two-dimensional vector space some pairs of $\mathbf{w}_1$, $\mathbf{w}_2$, $\mathbf{w}_3$, $\mathbf{w}_4$ form a non-obtuse angle. We assume that $\mathbf{w}_1$, $\mathbf{w}_2$ is such a pair, so that $0 \le \mathbf{w}_1^T\mathbf{w}_2 = w_{11}w_{21} + w_{12}w_{22}$. It follows that

$$\mathbf{v}_1^T\mathbf{v}_2 = v_{11}v_{21} + v_{12}v_{22} + v_{13}v_{23} > 0$$

and the angle θ_{12} between $\mathbf{v}_1$ and $\mathbf{v}_2$ is acute, i.e. $\theta_{12} \in (0, \pi/2)$. Thus we have a contradiction. Thus not all the angles formed by pairs can be obtuse.

Problem 3.   Let $\mathbf{k}$ be a normalized vector in $\mathbb{R}^3$. Let $\mathbf{x}$ be a vector in $\mathbb{R}^3$. Is

$$\mathbf{x} \equiv (\mathbf{k}\cdot\mathbf{x})\mathbf{k} + \mathbf{k}\times(\mathbf{x}\times\mathbf{k})\ ?$$

Here $\mathbf{k}\cdot\mathbf{x} = k_1x_1 + k_2x_2 + k_3x_3$ is the *scalar product* and $\times$ denotes the *vector product*.

Solution 3.   We have

$$
\mathbf{k} \times (\mathbf{x} \times \mathbf{k}) = \begin{pmatrix} x_1(k_2^2 + k_3^2) - k_1(k_2x_2 + k_3x_3) \\ x_2(k_1^2 + k_3^2) - k_2(k_1x_1 + k_3x_3) \\ x_3(k_1^2 + k_2^2) - k_3(k_1x_1 + k_2x_2) \end{pmatrix}
$$

and

$$
(\mathbf{k} \cdot \mathbf{x})\mathbf{k} = \begin{pmatrix} k_1(k_1x_1 + k_2x_2 + k_3x_3) \\ k_2(k_1x_1 + k_2x_2 + k_3x_3) \\ k_3(k_1x_1 + k_2x_2 + k_3x_3) \end{pmatrix}.
$$

Since $\|\mathbf{k}\| = \sqrt{k_1^2 + k_2^2 + k_3^2} = 1$ we find that the identity is true.

Problem 4.   Let

$$
\mathbf{a}_1 = \frac{a}{2}\begin{pmatrix} 1 & 1 & -1 \end{pmatrix}, \quad \mathbf{a}_2 = \frac{a}{2}\begin{pmatrix} -1 & 1 & 1 \end{pmatrix}, \quad \mathbf{a}_3 = \frac{a}{2}\begin{pmatrix} 1 & -1 & 1 \end{pmatrix} \quad (1)
$$

where a is a positive constant. Find the vectors $\mathbf{b}_j$ ($j = 1, 2, 3$) such that

$$
\mathbf{b}_j \cdot \mathbf{a}_k = 2\pi\delta_{jk} \tag{2}
$$

where $\mathbf{b}_j \cdot \mathbf{a}_k$ denotes the scalar product of $\mathbf{b}_j$ and $\mathbf{a}_k$, i.e.

$$
\mathbf{b}_j \cdot \mathbf{a}_k := \sum_{n=1}^{3} b_{jn}a_{kn}
$$

and

$$
\delta_{jk} := \begin{cases} 1 & j = k \\ 0 & \text{otherwise} \end{cases}.
$$

The vectors $\{\mathbf{a}_j : j = 1, 2, 3\}$ are a basis of the *body-centered cubic lattice*. The vectors $\{\mathbf{b}_j : j = 1, 2, 3\}$ are called the *reciprocal basis*.

Solution 4.   In matrix notation (2) can be written as

$$
BA = 2\pi I_3
$$

where B is the matrix whose rows are the vectors $\mathbf{b}_j$, A is the matrix whose columns are the vectors $\mathbf{a}_j$ and I_3 is the 3×3 unit matrix. Therefore

$$
A = \frac{a}{2}\begin{pmatrix} 1 & -1 & 1 \\ 1 & 1 & -1 \\ -1 & 1 & 1 \end{pmatrix}.
$$

We find $\det A \neq 0$. Consequently $B = 2\pi A^{-1}$. Using Gaussian elimination we find

$$
B = \frac{2\pi}{a}\begin{pmatrix} 1 & 1 & 0 \\ 0 & 1 & 1 \\ 1 & 0 & 1 \end{pmatrix}.
$$

Finally we arrive at

$$\mathbf{b}_1 = \frac{2\pi}{a}(1 \quad 1 \quad 0), \quad \mathbf{b}_2 = \frac{2\pi}{a}(0 \quad 1 \quad 1), \quad \mathbf{b}_3 = \frac{2\pi}{a}(1 \quad 0 \quad 1).$$

The vectors

$$\frac{a}{2}(1 \quad 1 \quad 0), \quad \frac{a}{2}(0 \quad 1 \quad 1), \quad \frac{a}{2}(1 \quad 0 \quad 1)$$

are a basis of the *face-centered cubic lattice*.

Problem 5.   Let A be a linear transformation of a vector space V into itself. Suppose $\mathbf{x} \in V$ is such that

$$A^n \mathbf{x} = \mathbf{0}, \qquad A^{n-1} \mathbf{x} \neq \mathbf{0}$$

for some positive integer n. Show that the vectors

$$\mathbf{x}, \; A\mathbf{x}, \; \ldots, \; A^{n-1}\mathbf{x}$$

are linearly independent.

Solution 5.   Suppose that there are scalars $c_0, c_1, \ldots, c_{n-1}$ such that

$$c_0 \mathbf{x} + c_1 A\mathbf{x} + \cdots + c_k A^k \mathbf{x} + \cdots + c_{n-1} A^{n-1}\mathbf{x} = \mathbf{0}.$$

Applying A^{n-1} to both sides, we obtain

$$c_0 A^{n-1}\mathbf{x} + c_1 A^n \mathbf{x} + \cdots + c_k A^{n-1+k}\mathbf{x} + \cdots + c_{n-1} A^{n-1+n-1}\mathbf{x} = \mathbf{0}$$

where we used that $A\mathbf{0} = \mathbf{0}$. Thus

$$c_0 A^{n-1}\mathbf{x} = \mathbf{0}$$

and therefore $c_0 = 0$. By the *induction principle* (multiplying by A^{n-k-1}) we find that all $c_k = 0$. Thus the set is linearly independent.

Problem 6.   Find all 2×2 matrices A with

$$\det(A) = a_{11}a_{22} - a_{12}a_{21} = 1 \tag{1}$$

and

$$A\frac{1}{\sqrt{2}}\begin{pmatrix} 1 \\ 1 \end{pmatrix} = \frac{1}{\sqrt{2}}\begin{pmatrix} 1 \\ 1 \end{pmatrix}. \tag{2}$$

Solution 6.   From (2) we find the system of linear equations

$$a_{11} + a_{12} = 1, \qquad a_{21} + a_{22} = 1.$$

From (1) we have $a_{11}a_{22} - a_{12}a_{21} = 1$. Eliminating a_{22} we obtain

$$a_{11} + a_{12} = 1, \qquad a_{11}(1 - a_{21}) - a_{12}a_{21} = 1.$$

Eliminating a_{12} we obtain $a_{11} - a_{21} = 1$ with $a_{12} = 1 - a_{11}$ and $a_{22} = 2 - a_{11}$. Thus

$$A = \begin{pmatrix} a_{11} & 1 - a_{11} \\ a_{11} - 1 & 2 - a_{11} \end{pmatrix}$$

with a_{11} arbitrary.

Problem 7. Let

$$\mathbf{x} = \begin{pmatrix} x_1 \\ x_2 \\ x_3 \end{pmatrix}, \qquad \mathbf{y} = \begin{pmatrix} y_1 \\ y_2 \\ y_3 \end{pmatrix}$$

be two normalized vectors in $\mathbb{R}^3$. Assume that $\mathbf{x}^T\mathbf{y} = 0$, i.e. the vectors are orthogonal. Is the vector $\mathbf{x} \times \mathbf{y}$ a unit vector again?

Solution 7. We have

$$\mathbf{x} \times \mathbf{y} = \begin{pmatrix} x_2y_3 - x_3y_2 \\ x_3y_1 - x_1y_3 \\ x_1y_2 - x_2y_1 \end{pmatrix} \tag{1}$$

$$x_1^2 + x_2^2 + x_3^2 = 1, \qquad y_1^2 + y_2^2 + y_3^2 = 1 \tag{2}$$

and

$$\mathbf{x}^T\mathbf{y} = x_1y_1 + x_2y_2 + x_3y_3 = 0. \tag{3}$$

Now the square of the norm of the vector $\mathbf{x} \times \mathbf{y}$ is given by

$$\|\mathbf{x} \times \mathbf{y}\|^2 = x_2^2y_3^2 + x_3^2y_2^2 + x_3^2y_1^2 + x_1^2y_3^2 + x_1^2y_2^2 + x_2^2y_1^2$$
$$- 2(x_2x_3y_2y_3 + x_1x_3y_1y_3 + x_1x_2y_1y_2).$$

From (2) it follows that

$$x_3^2 = 1 - x_1^2 - x_2^2, \quad y_3^2 = 1 - y_1^2 - y_2^2.$$

From (3) it follows that

$$x_3y_3 = -x_1y_1 - x_2y_2.$$

By rearranging and squaring this equation we obtain

$$2x_1y_1x_2y_2 = x_3^2y_3^2 - x_1^2y_1^2 - x_2^2y_2^2$$
$$= 1 - y_1^2 - y_2^2 - x_1^2 - x_2^2 + x_1^2y_2^2 + x_2^2y_1^2.$$

Inserting these equations into $\|\mathbf{x} \times \mathbf{y}\|^2$ gives $\|\mathbf{x} \times \mathbf{y}\|^2 = 1$. Thus the vector $\mathbf{x} \times \mathbf{y}$ is normalized.

Problem 8.   (i) Find four vectors $\mathbf{a}_1$, $\mathbf{a}_2$, $\mathbf{a}_3$, $\mathbf{a}_4$ in $\mathbb{R}^3$ such that

$$\mathbf{a}_j^T \mathbf{a}_k = \frac{4}{3}\delta_{jk} - \frac{1}{3} = \begin{cases} 1 & \text{for } j = k \\ -1/3 & \text{for } j \neq k \end{cases}.$$

(ii) Calculate

$$\sum_{j=1}^{4} \mathbf{a}_j, \qquad \frac{3}{4}\sum_{j=1}^{4} \mathbf{a}_j \mathbf{a}_j^T.$$

Solution 8.   (i) Geometrically speaking, such a quartet of vectors consists of the vectors pointing from the center of a cube to nonadjacent corners. Alternatively, we may picture these four vectors as the normal vectors for the faces of the *tetrahedron* that is defined by the other four corners of the cube. Owing to the conditions the four vectors are normalized. Thus

$$\mathbf{a}_1 = \frac{1}{\sqrt{3}}\begin{pmatrix} 1 \\ 1 \\ 1 \end{pmatrix}, \quad \mathbf{a}_2 = \frac{1}{\sqrt{3}}\begin{pmatrix} 1 \\ -1 \\ -1 \end{pmatrix}, \quad \mathbf{a}_3 = \frac{1}{\sqrt{3}}\begin{pmatrix} -1 \\ 1 \\ -1 \end{pmatrix}, \quad \mathbf{a}_4 = \frac{1}{\sqrt{3}}\begin{pmatrix} -1 \\ -1 \\ 1 \end{pmatrix}.$$

(ii) We have

$$\sum_{j=1}^{4} \mathbf{a}_j = \mathbf{0}, \qquad \frac{3}{4}\sum_{j=1}^{4} \mathbf{a}_j \mathbf{a}_j^T = I_3.$$

Problem 9.   Consider the normalized vector $\mathbf{v}_0 = \begin{pmatrix} 1 & 0 & 0 \end{pmatrix}^T$ in $\mathbb{R}^3$. Find three normalized vectors $\mathbf{v}_1$, $\mathbf{v}_2$, $\mathbf{v}_3$ such that

$$\sum_{j=0}^{3} \mathbf{v}_j = \mathbf{0}, \qquad \mathbf{v}_j^T \mathbf{v}_k = -\frac{1}{3} \quad (j \neq k).$$

Solution 9.   Owing to the form of $\mathbf{v}_0$ we have due to the second condition that $v_{1,1} = v_{2,1} = v_{3,1} = -1/3$. We obtain

$$\mathbf{v}_1 = \begin{pmatrix} -1/3 \\ 2\sqrt{2}/3 \\ 0 \end{pmatrix}, \qquad \mathbf{v}_2 = \begin{pmatrix} -1/3 \\ -\sqrt{2}/3 \\ \sqrt{6}/3 \end{pmatrix}, \qquad \mathbf{v}_3 = \begin{pmatrix} -1/3 \\ -\sqrt{2}/3 \\ -\sqrt{6}/3 \end{pmatrix}.$$

Problem 10.   Find the set of all four (column) vectors $\mathbf{u}_1$, $\mathbf{u}_2$, $\mathbf{v}_1$, $\mathbf{v}_2$ in $\mathbb{R}^2$ such that the following conditions are satisfied

$$\mathbf{v}_1^T \mathbf{u}_2 = 0, \quad \mathbf{v}_2^T \mathbf{u}_1 = 0, \quad \mathbf{v}_1^T \mathbf{u}_1 = 1, \quad \mathbf{v}_2^T \mathbf{u}_2 = 1.$$

Solution 10.   We obtain the following four conditions

$$v_{11}u_{21} + v_{12}u_{22} = 0, \qquad v_{21}u_{11} + v_{22}u_{12} = 0$$

$$v_{11}u_{11} + v_{12}u_{12} = 1, \qquad v_{21}u_{21} + v_{22}u_{22} = 1.$$

We have four equations with eight unkowns. A possible solution is

$$\mathbf{u}_1 = \begin{pmatrix} 0.86 \\ -0.52 \end{pmatrix}, \quad \mathbf{u}_2 = \begin{pmatrix} 0.92 \\ 0.40 \end{pmatrix}, \quad \mathbf{v}_1 = \begin{pmatrix} 0.49 \\ -1.12 \end{pmatrix}, \quad \mathbf{v}_2 = \begin{pmatrix} 0.63 \\ 1.04 \end{pmatrix}.$$

Problem 11.   Let x^μ be the standard coordinates in $\mathbb{R}^4$ with $\mu = 0, 1, 2, 3$. We define new coordinates x^{PQ} $(P, Q = 1, 2)$ by

$$\begin{pmatrix} x^{11} & x^{12} \\ x^{21} & x^{22} \end{pmatrix} := \begin{pmatrix} x^0 - ix^3 & -ix^1 - x^2 \\ -ix^1 + x^2 & x^0 + ix^3 \end{pmatrix}.$$

Let $Z^\alpha = (\omega^P, \pi_Q)$ be four complex coordinates on *twistor space* $\mathbb{C}^4$, i.e.

$$Z^0 = \omega^1, \quad Z^1 = \omega^2, \quad Z^2 = \pi_1, \quad Z^3 = \pi_2.$$

The basic equation expressing the relationship between Z^α and x^μ is (summation convention)

$$\omega^P = x^{PQ}\pi_Q. \tag{1}$$

The reality structure on Z^α is given by the antilinear map

$$Z^\alpha \mapsto Z^{*\alpha} = (\overline{Z^1}, -\overline{Z^0}, \overline{Z^3}, -\overline{Z^2})$$
$$\omega^P \mapsto \omega^{*P} = (\overline{\omega^2}, -\overline{\omega^1})$$
$$\pi_P \mapsto \pi_P^* = (\overline{\pi_2}, -\overline{\pi_1}).$$

Show that equation (1) is preserved under this map, in the sense that $\omega^P = x^{PQ}\pi_Q$ if and only if

$$\omega^{*P} = x^{PQ}\pi_Q^*.$$

Solution 11.   We have $x^{11} = \overline{x^{22}}$ and $\overline{x^{12}} = -x^{21}$. From $\omega^P = x^{PQ}\pi_Q$ we have

$$\omega^1 = x^{11}\pi_1 + x^{12}\pi_2, \qquad \omega^2 = x^{21}\pi_1 + x^{22}\pi_2$$

and from $\omega^{*P} = x^{PQ}\pi_Q^*$ we have

$$\omega^{*1} = x^{11}\pi_1^* + x^{12}\pi_2^*, \qquad \omega^{*2} = x^{21}\pi_1^* + x^{22}\pi_2^*.$$

Applying the antilinear map we have

$$\omega^{*1} = x^{11}\pi_1^* + x^{12}\pi_2^*$$
$$\Leftrightarrow \overline{\omega^2} = x^{11}\overline{\pi_2} - x^{12}\overline{\pi_1}$$
$$\Leftrightarrow \omega^2 = \overline{x^{11}}\pi_2 - \overline{x^{12}}\pi_1$$
$$\Leftrightarrow \omega^2 = x^{22}\pi_2 + x^{21}\pi_1.$$

Analogously we show that from $\omega^{*2} = x^{21}\pi_1^* + x^{22}\pi_2^*$ follows $\omega^1 = x^{12}\pi_2 + x^{11}\pi_1$ and that $\omega^P = x^{PQ}\pi_Q$ follows from $\omega^{*P} = x^{PQ}\pi_Q^*$.

Problem 12.   (i) Let R be a *nonsingular* $n \times n$ matrix (i.e. R^{-1} exists). Let A and B be two arbitrary $n \times n$ matrices. Assume that $R^{-1}AR$ and $R^{-1}BR$ are diagonal matrices. Show that

$$[A, B] = 0$$

where $[A, B]$ denotes the *commutator*, i.e. $[A, B] := AB - BA$.
(ii) Let X be an arbitrary $n \times n$ matrix. Let U be a nonsingular $n \times n$ matrix. Assume that $UXU^{-1} = X$. Show that $[X, U] = 0$.

Solution 12.   (i) Since $R^{-1}AR$ and $R^{-1}BR$ are diagonal matrices it follows that

$$[R^{-1}AR, R^{-1}BR] = 0.$$

This means that $R^{-1}AR$ and $R^{-1}BR$ commute. Therefore

$$R^{-1}ARR^{-1}BR - R^{-1}BRR^{-1}AR = 0.$$

It follows that

$$R^{-1}ABR - R^{-1}BAR = R^{-1}[A, B]R = 0.$$

Finally

$$RR^{-1}[A, B]RR^{-1} = [A, B] = 0.$$

(ii) From $UXU^{-1} = X$ we obtain $UXU^{-1} - X = 0$. Multiplying from the right with U, we find $UX - XU \equiv [U, X] = 0$.

Problem 13.   Let A and B be two $n \times n$ matrices. Assume that B is nonsingular. This means that $\det B \neq 0$. Therefore the inverse of B exists. Show that

$$[A, B^{-1}] = -B^{-1}[A, B]B^{-1}$$

where $[\,,\,]$ denotes the commutator.

Solution 13. We have

$$[A, B^{-1}] \equiv AB^{-1} - B^{-1}A.$$

Since $BB^{-1} = I_n$ and $B^{-1}B = I_n$, where I_n is the $n \times n$ unit matrix, we can write

$$[A, B^{-1}] = AB^{-1} - B^{-1}A = B^{-1}BAB^{-1} - B^{-1}ABB^{-1}.$$

Thus

$$[A, B^{-1}] = -B^{-1}(AB - BA)B^{-1} = -B^{-1}[A, B]B^{-1}\,.$$

Problem 14. Let $a, b, c \in \mathbb{R}$ and let

$$A(a,b,c) = \begin{pmatrix} a & b & 0 & 0 & \cdots & 0 & 0 & 0 \\ c & a & b & 0 & \cdots & 0 & 0 & 0 \\ 0 & c & a & b & \cdots & 0 & 0 & 0 \\ & & \ddots & & \ddots & & & \\ \vdots & & & \ddots & & \ddots & & \\ & & & & \ddots & & \ddots & \\ 0 & 0 & 0 & 0 & \cdots & c & a & b \\ 0 & 0 & 0 & 0 & \cdots & 0 & c & a \end{pmatrix}$$

be an $n \times n$ tridiagonal matrix. In other words

$$A_{jk} = \begin{cases} b & \text{if } j = k+1 \\ c & \text{if } j = k-1 \\ a & \text{if } j = k \\ 0 & \text{otherwise} \end{cases}$$

with $j, k = 1, 2, \ldots, n$. Calculate $\det A(a, b, c)$.

Solution 14. Let

$$D_n(a, b, c) := \det A(a, b, c).$$

By expanding in minors on the first row, we find the linear second-order difference equation with constant coefficients

$$D_{n+2}(a, b, c) = aD_{n+1}(a, b, c) - bcD_n(a, b, c) \tag{1}$$

where $n = 1, 2, \ldots$. Obviously the initial values of this difference equation are given by

$$D_1(a, b, c) = a, \qquad D_2(a, b, c) = a^2 - bc.$$

Since the linear difference equation (1) has constant coefficients we can solve it with the ansatz

$$D_n(a, b, c) = kr^n$$

where k is a constant, or we may employ the recurrence relations for the *Chebyshev polynomials* to obtain

$$D_n(a, b, c) = \begin{cases} (bc)^{n/2} U_n \left(\dfrac{a}{2\sqrt{bc}} \right) & \text{if} \quad bc \neq 0 \\ a^n & \text{if} \quad bc = 0 \end{cases}$$

where U_n is the n-th degree Chebyshev polynomial of the second kind given by

$$U_n(s) := \frac{(s + \sqrt{s^2 - 1})^{n+1} - (s - \sqrt{s^2 - 1})^{n+1}}{2\sqrt{s^2 - 1}}.$$

We notice that

$$a^n = \lim_{bc \to 0} (bc)^{n/2} U_n \left(\frac{a}{2\sqrt{bc}} \right).$$

The solution can also be given as

$$D_n(a, b, c) = a^n \prod_{j=1}^{n} \left(1 - \frac{2\sqrt{bc}}{a} \cos \left(\frac{j\pi}{n+1} \right) \right).$$

The ansatz $D_n(a, b, c) = kr^n$ leads to $r^2 = ar - bc$ with the solution

$$r_{1,2} = \frac{a}{2} \pm \sqrt{\frac{a^2}{4} - bc}.$$

Problem 15. Let C be an $n \times n$ matrix. The *trace* of C is defined as

$$\mathrm{tr} C := \sum_{j=1}^{n} c_{jj} .$$

(i) Let A, B be two $n \times n$ matrices. Assume that

$$\mathrm{tr} A = 0, \qquad \mathrm{tr} B = 0 .$$

Can we conclude that $\mathrm{tr}(AB) = 0$?

(ii) Let A and B be two $n \times n$ matrices. Prove that

$$\mathrm{tr}(AB) \neq (\mathrm{tr} A)(\mathrm{tr} B)$$

in general.

Solution 15.   (i) The answer is no. Let

$$A = B = \begin{pmatrix} 0 & 1 \\ 1 & 0 \end{pmatrix}.$$

Then $\mathrm{tr}A = \mathrm{tr}B = 0$. However, since

$$AB = \begin{pmatrix} 1 & 0 \\ 0 & 1 \end{pmatrix}$$

we have $\mathrm{tr}(AB) = 2$.

(ii) As a counterexample we can consider the matrices given in (i).

Problem 16.   Let A be an $n \times n$ *hermitian matrix*, i.e.

$$A = A^* \equiv \bar{A}^T \tag{1}$$

where $\bar{}$ denotes the complex conjugate and ^{T} the transpose.

(i) Show that $A + iI_n$ is invertible, where I_n denotes the $n \times n$ unit matrix.

(ii) Show that

$$U := (A - iI_n)(A + iI_n)^{-1}$$

is a unitary matrix.

Solution 16.   (i) Since A is hermitian we can find a unitary matrix V such that VAV^* is a diagonal matrix, where $V^* = V^{-1}$. The diagonal elements are real (they are the eigenvalues of A). Thus

$$V(A + iI_n)V^* = VAV^* + iVI_nV^* = VAV^* + iI_n$$
$$= \mathrm{diag}(\lambda_1 + i, \lambda_2 + i, \dots, \lambda_n + i).$$

The inverse of this diagonal matrix exists and therefore the inverse of $A + iI_n$ exists since

$$\det(XYZ) \equiv \det(X)\det(Y)\det(Z)$$

for $n \times n$ matrices X, Y, Z.

(ii) Recall that U is a *unitary matrix* if $U^*U = I_n$, where $U^* \equiv \bar{U}^T$. Note that $A - iI_n$ and $A + iI_n$ commute with each other, i.e.,

$$[A - iI_n, A + iI_n] = 0.$$

Now

$$U^* = [(A+iI_n)^{-1}]^*(A-iI_n)^* = [(A+iI_n)^*]^{-1}(A+iI_n) = (A-iI_n)^{-1}(A+iI_n).$$

Consequently,

$$U^*U = (A - iI_n)^{-1}(A + iI_n)(A - iI_n)(A + iI_n)^{-1}$$
$$= (A - iI_n)^{-1}(A - iI_n)(A + iI_n)(A + iI_n)^{-1}$$
$$= I_n.$$

Thus U is a unitary matrix.

Problem 17.   Let X, Y and Z be arbitrary $n \times n$ matrices.
(i) Show that
$$[X, Y + Z] = [X, Y] + [X, Z].$$

(ii) Show that
$$[X, Y] = -[Y, X].$$

(iii) Calculate
$$[X, [Y, Z]] + [Z, [X, Y]] + [Y, [Z, X]]$$
where $[\,,\,]$ denotes the commutator, i.e., $[X, Y] := XY - YX$.

Solution 17.   (i) Straightforward calculation yields
$$[X, Y+Z] = X(Y+Z)-(Y+Z)X = XY-YX+XZ-ZX = [X, Y]+[X, Z].$$

(ii) Straightforward calculation yields
$$[X, Y] = XY - YX = -(YX - XY) = -[Y, X].$$

(iii) Since
$$[X, [Y, Z]] = [X, YZ - ZY] = XYZ - YZX - XZY + ZYX$$
$$[Z, [X, Y]] = [Z, XY - YX] = ZXY - XYZ - ZYX + YXZ$$
$$[Y, [Z, X]] = [Y, ZX - XZ] = YZX - ZXY - YXZ + XZY$$
we obtain, by adding these equations
$$[X, [Y, Z]] + [Z, [X, Y]] + [Y, [Z, X]] = 0.$$

This equation is called the *Jacobi identity*. The $n \times n$ matrices over the real
or complex numbers form a *Lie algebra* under the commutator. The $n \times n$
matrices over the real or complex numbers form an *associative algebra* with
unit element (the unit matrix) under matrix multiplication.

Problem 18.   Let A, B and H be three $n \times n$ matrices. Assume that
$$[A, H] = 0, \qquad [B, H] = 0.$$

Calculate
$$[[A, B], H].$$

Solution 18.   For arbitrary $n \times n$ matrices X, Y and Z we have (Jacobi
identity)
$$[X, [Y, Z]] + [Z, [X, Y]] + [Y, [Z, X]] = 0.$$

Now let $X \equiv A$, $Y \equiv B$ and $Z \equiv H$. Since $[A, H] = 0$ and $[B, H] = 0$ it follows that

$$[H, [A, B]] = 0.$$

Since

$$[H, [A, B]] = -[[A, B], H]$$

we obtain

$$[[A, B], H] = 0.$$

If we assume

$$[H, A] = A, \qquad [B, H] = B$$

we also find

$$[[A, B], H] = 0.$$

Problem 19.   Let A and B be two $n \times n$ matrices. Assume that A^{-1} exists. Find the expansion of

$$(A - \epsilon B)^{-1}$$

as a power series in ϵ, where ϵ is a real parameter. Set

$$(A - \epsilon B)^{-1} := \sum_{n=0}^{\infty} \epsilon^n L_n. \qquad (1)$$

Solution 19.   Multiplying (1) on the left by $A - \epsilon B$ we obtain

$$I_n = \sum_{n=0}^{\infty} \epsilon^n (A - \epsilon B) L_n = AL_0 + \sum_{n=1}^{\infty} \epsilon^n (AL_n - BL_{n-1})$$

where I_n is the $n \times n$ unit matrix. By equating coefficients of powers of ϵ we find that $L_0 = A^{-1}$ and

$$L_n = A^{-1} B L_{n-1}$$

where $n = 1, 2, \ldots$. Consequently,

$$(A - \epsilon B)^{-1} = A^{-1} + \epsilon A^{-1} B A^{-1} + \epsilon^2 A^{-1} B A^{-1} B A^{-1} + \cdots. \qquad (2)$$

If $A = I_n$, then expansion (2) takes the form

$$(I_n - \epsilon B)^{-1} = I_n + \epsilon B + \epsilon^2 B^2 + \cdots.$$

Discuss the radius of convergence of this series.

Problem 20.   (i) Let A and B be two hermitian matrices. Is the commutator $[A, B]$ again a hermitian matrix?
(ii) Let A and B be two skew-hermitian matrices. Is the commutator $[A, B]$ again a skew-hermitian matrix?

Solution 20.   (i) Since A and B are hermitian matrices we have $A^* = A$, $B^* = B$. Now

$$([A, B])^* = (AB - BA)^* = (AB)^* - (BA)^* = B^* A^* - A^* B^*.$$

Thus

$$([A, B])^* = BA - AB = [B, A].$$

Consequently, in general the commutator of two hermitian matrices is not a hermitian matrix, since

$$[A, B] \neq [B, A]$$

in general.
(ii) Since A and B are two *skew-hermitian matrices* we have $A^* = -A$, $B^* = -B$. Now

$$([A, B])^* = (AB - BA)^* = (AB)^* - (BA)^*.$$

Thus

$$([A, B])^* = B^* A^* - A^* B^* = BA - AB = -[A, B].$$

Consequently, the commutator of two skew-hermitian matrices is again a skew-hermitian matrix. Notice that

$$\text{tr}([X, Y]) = 0$$

for two arbitrary $n \times n$ matrices X and Y over $\mathbb{R}$ or $\mathbb{C}$.

Problem 21.   Let A, B and C be three arbitrary $n \times n$ matrices.
(i) Show that

$$\text{tr}(AB) = \text{tr}(BA).$$

(ii) Calculate

$$\text{tr}([A, B]).$$

(iii) Find 2×2 matrices A, B and C such that

$$\text{tr}(ABC) \neq \text{tr}(ACB).$$

Solution 21.   (i) Since

$$(AB)_{ij} = \sum_{k=1}^{n} a_{ik} b_{kj}$$

the diagonal elements of AB are given by

$$(AB)_{ii} = \sum_{k=1}^{n} a_{ik} b_{ki}.$$

Since

$$(BA)_{ij} = \sum_{k=1}^{n} b_{ik} a_{kj}$$

the diagonal elements of BA are given by

$$(BA)_{ii} = \sum_{k=1}^{n} b_{ik} a_{ki} = \sum_{k=1}^{n} a_{ki} b_{ik}.$$

Now

$$\text{tr}(AB) = (AB)_{11} + (AB)_{22} + \cdots + (AB)_{nn}$$

and

$$\text{tr}(BA) = (BA)_{11} + (BA)_{22} + \cdots + (BA)_{nn}$$

so that

$$\text{tr}(AB) = \sum_{i=1}^{n} \sum_{k=1}^{n} a_{ik} b_{ki}, \qquad \text{tr}(BA) = \sum_{i=1}^{n} \sum_{k=1}^{n} a_{ki} b_{ik}.$$

Therefore, identity (1) holds.

(ii) Using identity (1) we have

$$\text{tr}([A, B]) = \text{tr}(AB - BA) = \text{tr}(AB) - \text{tr}(BA) = 0.$$

(iii) Let

$$A = \begin{pmatrix} 1 & 0 \\ 0 & 0 \end{pmatrix}, \qquad B = \begin{pmatrix} 0 & 1 \\ 0 & 0 \end{pmatrix}, \qquad C = \begin{pmatrix} 0 & 0 \\ 1 & 0 \end{pmatrix}.$$

Then

$$ABC = \begin{pmatrix} 1 & 0 \\ 0 & 0 \end{pmatrix}, \qquad ACB = \begin{pmatrix} 0 & 0 \\ 0 & 0 \end{pmatrix}.$$

Consequently

$$\text{tr}(ABC) = 1, \qquad \text{tr}(ACB) = 0.$$

Problem 22. Let $A(\epsilon)$ be an $n \times n$ matrix which depends on a real parameter ϵ. Assume that $dA/d\epsilon$ and $A^{-1}(\epsilon)$ exist for all ϵ. Show that

$$\frac{dA^{-1}}{d\epsilon} = -A^{-1}(\epsilon) \frac{dA}{d\epsilon} A^{-1}(\epsilon). \tag{1}$$

Solution 22.   We start from the identity

$$A(\epsilon)A^{-1}(\epsilon) = I_n$$

where I_n is the $n \times n$ unit matrix. Taking the derivative of this identity with respect to ϵ yields

$$\frac{dA}{d\epsilon}A^{-1}(\epsilon) + A(\epsilon)\frac{dA^{-1}}{d\epsilon} = 0$$

since $dI_n/d\epsilon = 0$. Therefore

$$\frac{dA}{d\epsilon}A^{-1}(\epsilon) = -A(\epsilon)\frac{dA^{-1}}{d\epsilon}.$$

Multiplying both sides from the left by $A^{-1}(\epsilon)$ we find (1).

Problem 23.   Let A be an $n \times n$ matrix. The entries depend smoothly on a real parameter ϵ. Show that in general

$$A(\epsilon)\frac{dA}{d\epsilon} \neq \frac{dA}{d\epsilon}A(\epsilon). \tag{1}$$

Solution 23.   Let

$$A(\epsilon) := \begin{pmatrix} \epsilon & \epsilon^2 \\ \epsilon^3 & 1 \end{pmatrix}.$$

Then

$$\frac{dA}{d\epsilon} = \begin{pmatrix} 1 & 2\epsilon \\ 3\epsilon^2 & 0 \end{pmatrix}.$$

Thus

$$A\frac{dA}{d\epsilon} = \begin{pmatrix} \epsilon + 3\epsilon^4 & 2\epsilon^2 \\ 3\epsilon^2 + \epsilon^3 & 2\epsilon^4 \end{pmatrix}, \qquad \frac{dA}{d\epsilon}A = \begin{pmatrix} \epsilon + 2\epsilon^4 & 2\epsilon + \epsilon^2 \\ 3\epsilon^3 & 3\epsilon^4 \end{pmatrix}.$$

Thus (1) holds in general. In particular we have

$$\frac{d}{d\epsilon}A^2 = A\frac{dA}{d\epsilon} + \frac{dA}{d\epsilon}A \neq 2A\frac{dA}{d\epsilon}$$

in general.

Problem 24.   Let A be a square finite-dimensional matrix over $\mathbb{R}$ such that

$$AA^T = I_n. \tag{1}$$

Show that

$$A^T A = I_n. \tag{2}$$

Does (2) also hold for infinite dimensional matrices?

Solution 24. (i) Since $\det A = \det A^T$ and $\det I_n = 1$ we obtain from (1) that

$$(\det A)^2 = 1.$$

Therefore the inverse of A exists and we have $A^T = A^{-1}$ with $A^{-1}A = AA^{-1} = I_n$.

(ii) The answer is no. Let

$$A = \begin{pmatrix} 0 & 1 & 0 & 0 & 0 & \cdots \\ 0 & 0 & 1 & 0 & 0 & \cdots \\ 0 & 0 & 0 & 1 & 0 & \cdots \\ \vdots & \vdots & \vdots & \vdots & \ddots & \cdots \end{pmatrix}.$$

Then the transpose matrix A^T of A is given by

$$A^T = \begin{pmatrix} 0 & 0 & 0 & 0 & 0 & \cdots \\ 1 & 0 & 0 & 0 & 0 & \cdots \\ 0 & 1 & 0 & 0 & 0 & \cdots \\ 0 & 0 & 1 & 0 & 0 & \cdots \\ \vdots & \vdots & \vdots & \vdots & \ddots & \cdots \end{pmatrix}.$$

It follows that

$$AA^T = \operatorname{diag}(1,1,1,\ldots) \equiv I$$

and

$$A^T A = \operatorname{diag}(0,1,1,\ldots)$$

where I is the infinite-dimensional unit matrix. Hence $A^T A \neq AA^T$.

Problem 25. Consider the boundary value problem (Dirichlet boundary conditions)

$$\frac{d^2 u}{dx^2} + u = 0, \qquad u(0) = u(1) = 1$$

for the interval $[0,1]$. The exact solution is given by

$$u(x) = \cos(x) + \frac{1 - \cos(1)}{\sin(1)} \sin(x).$$

Find a matrix equation for the discretization

$$\frac{d^2 u}{dx^2} \longrightarrow \frac{u_{i-1} - 2u_i + u_{i+1}}{h^2}, \qquad u \longrightarrow u_i$$

where $h = 0.1$ and $u_i := u(i \cdot h)$.

Solution 25.   Since the interval is $[0, 1]$ and $h = 0.1$ we have $u_0, u_1, \ldots, u_{10}$, where u_0 and u_{10} are given by the boundary conditions. We have

$$i = 1: \quad u_0 - 2u_1 + u_2 + h^2 u_1 = 0$$
$$i = 2: \quad u_1 - 2u_2 + u_3 + h^2 u_2 = 0$$
$$i = 3: \quad u_2 - 2u_3 + u_4 + h^2 u_3 = 0$$
$$\vdots$$
$$i = 9: \quad u_8 - 2u_9 + u_{10} + h^2 u_9 = 0.$$

The first equation can be brought into the form

$$-2u_1 + u_2 + h^2 u_1 = -u_0.$$

The last equation can be brought into the form

$$u_8 - 2u_9 + h^2 u_9 = -u_{10}.$$

From the boundary conditions $u(0) = 1$ and $u(1) = 1$ we obtain

$$u_0 = 1, \qquad u_{10} = 1.$$

Thus we get the matrix form of the discretization with a 9×9 matrix

$$
\begin{pmatrix}
\alpha & 1 & 0 & 0 & 0 & 0 & 0 & 0 & 0 \\
1 & \alpha & 1 & 0 & 0 & 0 & 0 & 0 & 0 \\
0 & 1 & \alpha & 1 & 0 & 0 & 0 & 0 & 0 \\
0 & 0 & 1 & \alpha & 1 & 0 & 0 & 0 & 0 \\
0 & 0 & 0 & 1 & \alpha & 1 & 0 & 0 & 0 \\
0 & 0 & 0 & 0 & 1 & \alpha & 1 & 0 & 0 \\
0 & 0 & 0 & 0 & 0 & 1 & \alpha & 1 & 0 \\
0 & 0 & 0 & 0 & 0 & 0 & 1 & \alpha & 1 \\
0 & 0 & 0 & 0 & 0 & 0 & 0 & 1 & \alpha
\end{pmatrix}
\begin{pmatrix}
u_1 \\ u_2 \\ u_3 \\ u_4 \\ u_5 \\ u_6 \\ u_7 \\ u_8 \\ u_9
\end{pmatrix}
=
\begin{pmatrix}
-1 \\ 0 \\ 0 \\ 0 \\ 0 \\ 0 \\ 0 \\ 0 \\ -1
\end{pmatrix}
$$

with the abbreviation

$$\alpha := -2 + h^2.$$

The matrix on the left-hand side is a tridiagonal matrix. The inverse of the matrix exists. The solution algorithm for the tridiagonal equation is called the tridiagonal solution (a variant of Gaussian elimination).

Chapter 6

Matrices and Groups

A *group* G is a set of objects $\{a, b, c, \dots\}$ (not necessarily countable) together with a binary operation which associates with any ordered pair of elements a, b in G a third element ab in G (closure). The binary operation (called group multiplication) is subject to the following requirements:

(1) There exists an element e in G called the *identity element* such that $eg = ge = g$ for all $g \in G$.

(2) For every $g \in G$ there exists an *inverse element* g^{-1} in G such that $gg^{-1} = g^{-1}g = e$.

(3) *Associative law.* The identity $(ab)c = a(bc)$ is satisfied for all $a, b, c \in G$.

If $ab = ba$ for all $a, b \in G$ we call the group *commutative*. If G has a finite number of elements it has *finite order* $n(G)$, where $n(G)$ is the number of elements. Otherwise, G has infinite order.

Groups have matrix representations with invertible $n \times n$ matrices and matrix multiplication as group multiplication. The identity element is the identity matrix. The inverse element is the inverse matrix. An important subgroup is the set of unitary matrices, where $U^* = U^{-1}$.

Let $(G_1, *)$ and $(G_2, \circ)$ be groups. A function $f : G_1 \to G_2$ with

$$f(a * b) = f(a) \circ f(b), \qquad \text{for all } a, b \in G_1$$

is called a *homomorphism*.

Problem 1. Let

$$A = \begin{pmatrix} 1 & 0 & 0 \\ 0 & 1 & 0 \\ 0 & 0 & 1 \end{pmatrix}, \quad B = \begin{pmatrix} 1 & 0 & 0 \\ 0 & 0 & 1 \\ 0 & 1 & 0 \end{pmatrix}, \quad C = \begin{pmatrix} 0 & 1 & 0 \\ 1 & 0 & 0 \\ 0 & 0 & 1 \end{pmatrix},$$

$$D = \begin{pmatrix} 0 & 1 & 0 \\ 0 & 0 & 1 \\ 1 & 0 & 0 \end{pmatrix}, \quad E = \begin{pmatrix} 0 & 0 & 1 \\ 1 & 0 & 0 \\ 0 & 1 & 0 \end{pmatrix}, \quad F = \begin{pmatrix} 0 & 0 & 1 \\ 0 & 1 & 0 \\ 1 & 0 & 0 \end{pmatrix}.$$

(i) Show that these matrices form a group under matrix multiplication.
(ii) Find all subgroups.
These matrices are called *permutation matrices*. Why?

Solution 1. (i) Since

$$AA = A \quad AB = B \quad AC = C \quad AD = D \quad AE = E \quad AF = F$$
$$BA = B \quad BB = A \quad BC = D \quad BD = C \quad BE = F \quad BF = E$$
$$CA = C \quad CB = E \quad CC = A \quad CD = F \quad CE = B \quad CF = D$$
$$DA = D \quad DB = F \quad DC = B \quad DD = E \quad DE = A \quad DF = C$$
$$EA = E \quad EB = C \quad EC = F \quad ED = A \quad EE = D \quad EF = B$$
$$FA = F \quad FB = D \quad FC = E \quad FD = B \quad FE = C \quad FF = A$$

we find the following. The set of matrices given above is closed under matrix multiplication. The neutral element is the matrix A, i.e. the unit matrix. Each element has an inverse. From the table given above we find that

$$A^{-1} = A, \quad B^{-1} = B, \quad C^{-1} = C, \quad D^{-1} = E, \quad E^{-1} = D, \quad F^{-1} = F.$$

Since the associative law holds for matrices, we find that the matrices given above form a finite group under matrix multiplication.
(ii) The *order of a finite group* is the number of elements of the group. Thus our group has order 6. *Lagrange's theorem* tells us that the order of a subgroup of a finite group divides the order of the group. Thus the subgroups must have order 3, 2, 1. From the group table we find (besides the group itself) the subgroups

$$\{A, D, E\}$$

$$\{A, B\}, \quad \{A, C\}, \quad \{A, F\}$$

$$\{A\}$$

The set of all $n \times n$ permutation matrices form a group under matrix multiplication. *Cayley's theorem* tells us that every finite group is isomorphic to a subgroup (or the group itself) of these permutation matrices. The *order of an element* $g \in G$ is the order of the cyclic subgroup generated by $\{g\}$, i.e. the smallest positive integer m such that

$$g^m = e$$

where e is the identity element of the group. The integer m divides the order of G. Consider, for example, the element D of our group. Then

$$D^2 = E, \qquad D^3 = A, \qquad A \quad \text{identity element.}$$

Thus $m = 3$.

Problem 2.   Consider the set of Pauli spin matrices σ_x, σ_y, σ_z and the 2×2 identity matrix I_2. Can we extend this set so that we obtain a group under matrix multiplication?

Solution 2.   We have $\sigma_x^2 = \sigma_y^2 = \sigma_z^2 = I_2$ and

$$\sigma_x \sigma_y = i\sigma_z, \qquad \sigma_y \sigma_x = -i\sigma_z$$
$$\sigma_y \sigma_z = i\sigma_x, \qquad \sigma_z \sigma_y = -i\sigma_x$$
$$\sigma_z \sigma_x = i\sigma_y, \qquad \sigma_x \sigma_z = -i\sigma_y.$$

Thus we have to extend the set to the set of 16 elements

$$\{\, \pm I_2, \ \pm\sigma_x, \ \pm\sigma_y, \ \pm\sigma_z, \ \pm iI_2, \ \pm i\sigma_x, \ \pm i\sigma_y, \ \pm i\sigma_z \,\}$$

to obtain a group under matrix multiplication. Note that all matrices are unitary. A generating set is $\{\, \sigma_x, \sigma_z \,\}$.

Problem 3.   Let $GL(n, \mathbb{F})$ be the group of invertible $n \times n$ matrices with entries in the field $\mathbb{F}$, where $\mathbb{F}$ is $\mathbb{R}$ or $\mathbb{C}$. Let G be a group. A *matrix representation* of G over the field $\mathbb{F}$ is a homomorphism ρ from G to $GL(n, \mathbb{F})$. The degree of ρ is the integer n. Let $\rho : G \to GL(n, \mathbb{F})$. Then ρ is a reprentation if and only if

$$\rho(g \circ h) = \rho(g)\rho(h)$$

for all $g, h \in G$. Let G be the *dihedral group* defined by

$$D_8 := \langle a, b \ : \ a^4 = b^2 = 1, \ b^{-1}ab = a^{-1} \rangle$$

where 1 is the identity element in the group. Define the invertible 2×2 matrices

$$A = \begin{pmatrix} 0 & 1 \\ -1 & 0 \end{pmatrix}, \qquad B = \begin{pmatrix} 1 & 0 \\ 0 & -1 \end{pmatrix}.$$

Let I_2 be the 2×2 identity matrix. Show that

$$A^4 = B^2 = I_2, \qquad B^{-1}AB = A^{-1}$$

and thus show that we have a matrix representation of the dihedral group.

Solution 3.   We have $A^2 = -I_2$. Thus $A^4 = I_2$. Furthermore $B^2 = I_2$. The inverse of B is given by $B^{-1} = B$. The inverse of A is $-A$. Thus

$$B^{-1}AB = BAB = \begin{pmatrix} 1 & 0 \\ 0 & -1 \end{pmatrix} \begin{pmatrix} 0 & 1 \\ -1 & 0 \end{pmatrix} \begin{pmatrix} 1 & 0 \\ 0 & -1 \end{pmatrix} = A^{-1}.$$

It follows that the map

$$\rho : a^j b^k \to A^j B^k \qquad j = 0, 1, 2, 3 \quad k = 0, 1$$

provides a matrix representation of D_8 over $\mathbb{R}$.

Problem 4.   Consider the group $G = \{a, b, c\}$ with group operation $\cdot : G \times G \to G$ defined by the *group table*

$\cdot$	a	b	c
a	a	b	c
b	b	c	a
c	c	a	b

(i) Let $M(n)$ denote the vector space of $n \times n$ matrices over $\mathbb{C}$. Show that $f : G \to M(3)$ is a matrix representation for $(G, \cdot)$ where

$$f(a) := \begin{pmatrix} 1 & 0 & 0 \\ 0 & 1 & 0 \\ 0 & 0 & 1 \end{pmatrix} = I_3$$

$$f(b) := \begin{pmatrix} \dfrac{\sqrt{3}}{2}i - \dfrac{1}{2} & 0 & 0 \\ 0 & -\dfrac{1}{2} & \dfrac{\sqrt{3}}{2} \\ 0 & -\dfrac{\sqrt{3}}{2} & -\dfrac{1}{2} \end{pmatrix}$$

$$f(c) := \begin{pmatrix} -\dfrac{\sqrt{3}}{2}i - \dfrac{1}{2} & 0 & 0 \\ 0 & -\dfrac{1}{2} & -\dfrac{\sqrt{3}}{2} \\ 0 & \dfrac{\sqrt{3}}{2} & -\dfrac{1}{2} \end{pmatrix}.$$

(ii) Is the matrix representation in (i) reducible? Prove or disprove.

(iii) Let $f : G \to M(m)$ be a representation of a group $(G, \cdot)$ and let $g : G \to M(n)$ also be a representation of the same group $(G, \cdot)$. Show that the maps $h : G \to M(mn)$ and $k : G \to M(m+n)$ defined by

$$h(x) := f(x) \otimes g(x), \qquad k(x) := f(x) \oplus g(x)$$

for all $x \in G$, is also a group representation of $(G, \cdot)$. Here $\otimes$ denotes the Kronecker product of matrices and $\oplus$ denotes the direct sum of matrices.

Solution 4.   (i) We obtain

$$
\begin{aligned}
f(a)f(a) &= f(a), & f(a)f(b) &= f(b), & f(a)f(c) &= f(c) \\
f(b)f(a) &= f(b), & f(b)f(b) &= f(c), & f(b)f(c) &= f(a) \\
f(c)f(a) &= f(c), & f(c)f(b) &= f(a), & f(c)f(c) &= f(b)
\end{aligned}
$$

Thus f is a matrix representation of the group.
(ii) We can write all three matrices as direct sums

$$f(a) = (1) \oplus \begin{pmatrix} 1 & 0 \\ 0 & 1 \end{pmatrix}$$

$$f(b) = (e^{2\pi i/3}) \oplus \begin{pmatrix} \cos\left(\dfrac{2\pi i}{3}\right) & \sin\left(\dfrac{2\pi i}{3}\right) \\ -\sin\left(\dfrac{2\pi i}{3}\right) & \cos\left(\dfrac{2\pi i}{3}\right) \end{pmatrix}$$

$$f(c) = (e^{4\pi i/3}) \oplus \begin{pmatrix} \cos\left(\dfrac{4\pi i}{3}\right) & \sin\left(\dfrac{4\pi i}{3}\right) \\ -\sin\left(\dfrac{4\pi i}{3}\right) & \cos\left(\dfrac{4\pi i}{3}\right) \end{pmatrix}.$$

Thus we find the invariant vector subspace

$$\left\{ \begin{pmatrix} t \\ 0 \\ 0 \end{pmatrix} \middle| t \in \mathbb{C} \right\}.$$

Consequently f is reducible. Another invariant vector subspace is

$$\left\{ \begin{pmatrix} 0 \\ u \\ v \end{pmatrix} \middle| u, v \in \mathbb{C} \right\}.$$

(iii) Using the properties of the Kronecker product and the direct sum we obtain

$$
\begin{aligned}
h(a \cdot b) &= f(a \cdot b) \otimes g(a \cdot b) \\
&= (f(a)f(b)) \otimes (g(a)g(b)) \\
&= (f(a) \otimes g(a))(f(b) \otimes g(b)) \\
&= h(a)h(b)
\end{aligned}
$$

and

$$k(a \cdot b) = f(a \cdot b) \oplus g(a \cdot b)$$
$$= (f(a)f(b)) \oplus (g(a)g(b))$$
$$= (f(a) \oplus g(a))(f(b) \oplus g(b))$$
$$= k(a)k(b).$$

It follows that h and k are also matrix representations.

Problem 5.   Consider the set of all 2×2 matrices over the set of integers $\mathbb{Z}$ with determinant equal to 1. Show that these matrices form a group under matrix multiplication. This group is called $SL(2, \mathbb{Z})$.

Solution 5.   Let $A, B \in SL(2, \mathbb{Z})$. We have

$$AB = \begin{pmatrix} a_{11} & a_{12} \\ a_{21} & a_{22} \end{pmatrix} \begin{pmatrix} b_{11} & b_{12} \\ b_{21} & b_{22} \end{pmatrix} = \begin{pmatrix} a_{11}b_{11} + a_{12}b_{21} & a_{11}b_{12} + a_{12}b_{22} \\ a_{21}b_{11} + a_{22}b_{21} & a_{21}b_{12} + a_{22}b_{22} \end{pmatrix}.$$

Since $a_{ij}b_{kl} \in \mathbb{Z}$, and $\det A = a_{11}a_{22} - a_{12}a_{21} = 1$, $\det B = b_{11}b_{22} - b_{12}b_{21} = 1$ with $\det(AB) = \det(A)\det(B) = 1$ we obtain that $AB \in SL(2, \mathbb{Z})$. The neutral element (identity) is the 2×2 identity matrix. The inverse of A is given by

$$A^{-1} = \begin{pmatrix} a_{22} & -a_{12} \\ -a_{21} & a_{11} \end{pmatrix}$$

with $\det A^{-1} = a_{11}a_{22} - a_{12}a_{21} = 1$ and $-a_{12}, -a_{21} \in \mathbb{Z}$. Matrix multiplication is associative. Thus we have a group.

Problem 6.   The underlying field is the real numbers $\mathbb{R}$. Given any rotation matrix, $R \in SO(n)$. If R does not admit -1 as an eigenvalue, then there is a unique skew symmetric matrix, S, $(S^T = -S)$ so that

$$R = (I_n - S)(I_n + S)^{-1}.$$

The matrix R is called the *Cayley transform* of S. Let

$$S = \begin{pmatrix} 0 & 1 \\ -1 & 0 \end{pmatrix}.$$

Find R.

Solution 6.   We have

$$I_2 - S = \begin{pmatrix} 1 & -1 \\ 1 & 1 \end{pmatrix}, \qquad I_2 + S = \begin{pmatrix} 1 & 1 \\ -1 & 1 \end{pmatrix}$$

and
$$(I_2 + S)^{-1} = \frac{1}{2} \begin{pmatrix} 1 & -1 \\ 1 & 1 \end{pmatrix}.$$

It follows that
$$(I_2 - S)(I_2 + S)^{-1} = \begin{pmatrix} 1 & -1 \\ 1 & 1 \end{pmatrix} \frac{1}{2} \begin{pmatrix} 1 & -1 \\ 1 & 1 \end{pmatrix} = \begin{pmatrix} 0 & -1 \\ 1 & 0 \end{pmatrix} = R.$$

Thus $R = S^T$.

Problem 7. (i) Let
$$A(\alpha) := \begin{pmatrix} \cos \alpha & \sin \alpha \\ -\sin \alpha & \cos \alpha \end{pmatrix}$$

where $\alpha \in \mathbb{R}$. Show that the matrices $A(\alpha)$ form a group under matrix multiplication.
(ii) Let
$$X := \left. \frac{dA(\alpha)}{d\alpha} \right|_{\alpha=0}.$$

Find X. Calculate $e^{\alpha X}$.
(iii) Let
$$B(\alpha) := \begin{pmatrix} \cosh \alpha & \sinh \alpha \\ \sinh \alpha & \cosh \alpha \end{pmatrix}.$$

Show that the matrices $B(\alpha)$ form a group under matrix multiplication.

Solution 7. (i) Since
$$A(\alpha)A(\beta) = \begin{pmatrix} \cos(\alpha + \beta) & \sin(\alpha + \beta) \\ -\sin(\alpha + \beta) & \cos(\alpha + \beta) \end{pmatrix} \tag{1}$$

the set is closed under multiplication. To find (1) we have used the identities
$$\cos \alpha \cos \beta - \sin \alpha \sin \beta \equiv \cos(\alpha + \beta)$$
$$\sin \alpha \cos \beta + \cos \alpha \sin \beta \equiv \sin(\alpha + \beta).$$

For $\alpha = 0$ we obtain the neutral element of the group (i.e. the unit matrix)
$$A(\alpha = 0) = \begin{pmatrix} 1 & 0 \\ 0 & 1 \end{pmatrix} = I_2.$$

Since $\det A(\alpha) - 1$ the inverse exists and is given by
$$A^{-1}(\alpha) = A(-\alpha) = \begin{pmatrix} \cos \alpha & -\sin \alpha \\ \sin \alpha & \cos \alpha \end{pmatrix}.$$

For arbitrary $n \times n$ matrices the associative law holds. Consequently the matrices $A(\alpha)$ form a group. The group is called $SO(2)$.

(ii) We find

$$X = \begin{pmatrix} 0 & 1 \\ -1 & 0 \end{pmatrix}$$

owing to $\sin(0) = 0$ and $\cos(0) = 1$. Since

$$e^{\alpha X} := \sum_{k=0}^{\infty} \frac{(\alpha X)^k}{k!}$$

we find

$$e^{\alpha X} = A(\alpha)$$

because $X^2 = -I_2$. In physics X is called the *generator* of the Lie group $SO(2)$.

(iii) Since

$$B(\alpha)B(\beta) = \begin{pmatrix} \cosh(\alpha + \beta) & \sinh(\alpha + \beta) \\ \sinh(\alpha + \beta) & \cosh(\alpha + \beta) \end{pmatrix} \tag{2}$$

the set is closed under multiplication. To find (2) we have used the identities

$$\cosh \alpha \cosh \beta + \sinh \alpha \sinh \beta \equiv \cosh(\alpha + \beta)$$

$$\sinh \alpha \cosh \beta + \cosh \alpha \sinh \beta \equiv \sinh(\alpha + \beta).$$

For $\alpha = 0$ we obtain the neutral element

$$B(\alpha = 0) = \begin{pmatrix} 1 & 0 \\ 0 & 1 \end{pmatrix}.$$

Since $\det B(\alpha) = 1$ the inverse matrix exists and is given by

$$B^{-1}(\alpha) = B(-\alpha) = \begin{pmatrix} \cosh \alpha & -\sinh \alpha \\ -\sinh \alpha & \cosh \alpha \end{pmatrix}.$$

For arbitrary $n \times n$ matrices the associative law holds. Consequently, the matrices $B(\alpha)$ form a group. The group is called $SO(1,1)$.

Chapter 7

Matrices and Eigenvalue Problems

Let A be an $n \times n$ matrix over $\mathbb{C}$. The *eigenvalue equation* is given by

$$A\mathbf{x} = \lambda\mathbf{x}$$

with $\mathbf{x} \neq \mathbf{0}$. Here λ is called the eigenvalue and $\mathbf{x}$ is called the corresponding eigenvector. Thus for any eigenvector $\mathbf{x}$ of A, $A\mathbf{x}$ is a scalar multiple of $\mathbf{x}$.

Problem 1. (i) Show that the eigenvalues of a hermitian matrix are real.
(ii) Show that the eigenvalues of a skew-hermitian matrix are purely imaginary or zero.
(iii) Show that the eigenvalues λ_j of a unitary matrix satisfy $|\lambda_j| = 1$.

Solution 1. (i) Since A is hermitian we have

$$A^* = A \tag{1}$$

where $A^* \equiv \bar{A}^T$. Now we have the identity

$$(A\mathbf{x})^*\mathbf{x} \equiv \mathbf{x}^* A^*\mathbf{x} \equiv \mathbf{x}^*(A^*\mathbf{x}). \tag{2}$$

Inserting (1) and the eigenvalue equation into (2) gives

$$(\lambda\mathbf{x})^*\mathbf{x} = \mathbf{x}^*(\lambda\mathbf{x}).$$

Consequently

$$\bar{\lambda}(\mathbf{x}^*\mathbf{x}) = \lambda(\mathbf{x}^*\mathbf{x}).$$

Since $\mathbf{x}^*\mathbf{x} \neq 0$ we have $\bar{\lambda} = \lambda$. Therefore λ must be real.
(ii) Since A is skew-hermitian we have

$$A^* = -A.$$

Using this property we have

$$(A\mathbf{x})^*\mathbf{x} \equiv \mathbf{x}^* A^*\mathbf{x} \equiv \mathbf{x}^*(A^*\mathbf{x}) \equiv -\mathbf{x}^*(A\mathbf{x}).$$

Inserting the eigenvalue equation yields

$$\bar{\lambda}(\mathbf{x}^*\mathbf{x}) = -\lambda(\mathbf{x}^*\mathbf{x}).$$

Since $\mathbf{x}^*\mathbf{x} \neq 0$ we have $\bar{\lambda} = -\lambda$. Thus the eigenvalues are purely imaginary or zero.
(iii) Since U is a unitary matrix we have $U^* = U^{-1}$, where U^{-1} is the inverse of U. Let

$$U\mathbf{x} = \lambda\mathbf{x}$$

be the eigenvalue equation. It follows that $(U\mathbf{x})^* = (\lambda\mathbf{x})^*$ or

$$\mathbf{x}^*U^* = \mathbf{x}^*\bar{\lambda}.$$

Therefore

$$\mathbf{x}^*U^*U\mathbf{x} = \bar{\lambda}\lambda\mathbf{x}^*\mathbf{x}.$$

Since $U^*U = I$ we have

$$\mathbf{x}^*\mathbf{x} = \bar{\lambda}\lambda\mathbf{x}^*\mathbf{x}.$$

Since $\mathbf{x}^*\mathbf{x} \neq 0$ we have $\bar{\lambda}\lambda = 1$. Thus λ can be written as $\lambda = e^{i\alpha}$, $\alpha \in \mathbb{R}$.

Problem 2.   Let

$$A = \begin{pmatrix} 0 & i \\ -i & 0 \end{pmatrix}.$$

(i) Calculate the eigenvalues and eigenvectors of the matrix A.
(ii) Are the eigenvectors orthogonal ?
(iii) Do the eigenvectors form a basis in $\mathbb{C}^2$?
(iv) Find a unitary matrix U such that

$$U^*AU = \begin{pmatrix} \lambda_1 & 0 \\ 0 & \lambda_2 \end{pmatrix}.$$

Obviously, λ_1 and λ_2 are the eigenvalues of A.

Solution 2.   (i) The *eigenvalues* are determined by the equation

$$\det(A - \lambda I_2) = 0$$

where det denotes the determinant and I_2 is the 2×2 unit matrix. This equation leads to the quadratic equation $\lambda^2 - 1 = 0$. Thus we obtain the eigenvalues $\lambda_1 = 1$, $\lambda_2 = -1$. Notice that the matrix A is hermitian and thus the eigenvalues must be real. The eigenvectors are determined by solving the linear equation

$$A \begin{pmatrix} x_1 \\ x_2 \end{pmatrix} = \lambda \begin{pmatrix} x_1 \\ x_2 \end{pmatrix}.$$

Therefore the normalized eigenvector which belongs to λ_1 is given by

$$\frac{1}{\sqrt{2}} \begin{pmatrix} 1 \\ -i \end{pmatrix}.$$

The normalized eigenvector which belongs to λ_2 is given by

$$\frac{1}{\sqrt{2}} \begin{pmatrix} 1 \\ i \end{pmatrix}.$$

(ii) Two vectors $\mathbf{x}$ and $\mathbf{y}$ are called *orthogonal* if $\bar{\mathbf{x}}^T \mathbf{y} = 0$. The two eigenvectors are orthogonal, i.e.

$$\frac{1}{\sqrt{2}} \begin{pmatrix} 1 & i \end{pmatrix} \frac{1}{\sqrt{2}} \begin{pmatrix} 1 \\ i \end{pmatrix} = 0$$

(iii) They form an orthonormal basis in the vector space $\mathbb{C}^2$.
(iv) The normalized eigenvectors given above lead to the unitary matrix

$$U = \frac{1}{\sqrt{2}} \begin{pmatrix} 1 & 1 \\ -i & i \end{pmatrix}.$$

Consequently,

$$U^* = \frac{1}{\sqrt{2}} \begin{pmatrix} 1 & i \\ 1 & -i \end{pmatrix}$$

where $U^* = U^{-1}$. Then we obtain

$$U^* A U = \begin{pmatrix} 1 & 0 \\ 0 & -1 \end{pmatrix}.$$

Let A be an arbitrary $n \times n$ hermitian matrix. Assume that all eigenvalues of A are different. Then the eigenvectors are pairwise orthogonal.

Problem 3. Calculate the eigenvalues λ_j of the matrices

$$A = \begin{pmatrix} 0 & 0 & 1 \\ 0 & 1 & 0 \\ 1 & 0 & 0 \end{pmatrix}$$

and

$$B = \begin{pmatrix} 0 & 0 & 0 & 1 \\ 0 & 0 & 1 & 0 \\ 0 & 1 & 0 & 0 \\ 1 & 0 & 0 & 0 \end{pmatrix}.$$

Use the property that the matrices are symmetric and orthogonal. Moreover use the fact that the trace of an $n \times n$ matrix is equal to the sum of the eigenvalues of the matrix, i.e.

$$\mathrm{tr}X = \sum_{j=1}^{n} \lambda_j.$$

Solution 3.   Since the matrices A and B are symmetric over $\mathbb{R}$ it follows that the eigenvalues λ_j are real. From the property that the matrices are orthogonal it follows that the eigenvalues are ± 1. From

$$\mathrm{tr}A = \lambda_1 + \lambda_2 + \lambda_3 = 1$$

we find that the eigenvalues of the matrix A are given by $\{1, 1, -1\}$. Since

$$\mathrm{tr}B = \lambda_1 + \lambda_2 + \lambda_3 + \lambda_4 = 0$$

we find that the eigenvalues of B are given by $\{1, 1, -1, -1\}$.

Problem 4.   (i) Let A be the symmetric matrix

$$A = \begin{pmatrix} 5 & -2 & -4 \\ -2 & 2 & 2 \\ -4 & 2 & 5 \end{pmatrix}$$

over $\mathbb{R}$. Calculate the eigenvalues and eigenvectors of A. Are the eigenvectors orthogonal to each other? If not, try to find orthogonal eigenvectors using the *Gram-Schmidt algorithm*.
(ii) Let B be an arbitrary symmetric $n \times n$ matrix over $\mathbb{R}$. Show that the eigenvectors which belong to different eigenvalues are orthogonal.

Solution 4.   (i) Since the matrix A is symmetric we find that the eigenvalues are real. The eigenvalues are determined by

$$\det(A - \lambda I_3) = 0.$$

From this equation we obtain the characteristic polynomial

$$-\lambda^3 + 12\lambda^2 - 21\lambda + 10 = 0.$$

The eigenvalues are $\lambda_1 = 1$, $\lambda_2 = 1$, $\lambda_3 = 10$ with the corresponding eigenvectors

$$\mathbf{u}_1 = \begin{pmatrix} -1 \\ -2 \\ 0 \end{pmatrix}, \quad \mathbf{u}_2 = \begin{pmatrix} -1 \\ 0 \\ -1 \end{pmatrix}, \quad \mathbf{u}_3 = \begin{pmatrix} 2 \\ -1 \\ -2 \end{pmatrix}.$$

We find

$$(\mathbf{u}_1, \mathbf{u}_3) = 0, \qquad (\mathbf{u}_2, \mathbf{u}_3) = 0, \qquad (\mathbf{u}_1, \mathbf{u}_2) = 1$$

where $(,)$ denotes the scalar product, i.e.

$$(\mathbf{u}_j, \mathbf{u}_k) := \mathbf{u}_j{}^T \mathbf{u}_k$$

and T denotes the transpose. To apply the Gram-Schmidt algorithm we choose

$$\mathbf{u'}_1 = \mathbf{u}_1 \qquad \mathbf{u'}_2 = \mathbf{u}_2 + \alpha \mathbf{u}_1$$

such that

$$\alpha = -\frac{(\mathbf{u}_1, \mathbf{u}_2)}{(\mathbf{u}_1, \mathbf{u}_1)} = -\frac{1}{5}.$$

Consequently,

$$\mathbf{u'}_2 = \begin{pmatrix} -4/5 \\ 2/5 \\ -1 \end{pmatrix}.$$

The vectors $\mathbf{u}_1, \mathbf{u'}_2, \mathbf{u}_3$ are orthogonal.

(ii) From the eigenvalue equations

$$B\mathbf{u}_j = \lambda_j \mathbf{u}_j, \quad B\mathbf{u}_k = \lambda_k \mathbf{u}_k$$

we obtain

$$\mathbf{u}_k^T B \mathbf{u}_j = \lambda_j \mathbf{u}_k^T \mathbf{u}_j, \qquad \mathbf{u}_j^T B \mathbf{u}_k = \lambda_k \mathbf{u}_j^T \mathbf{u}_k.$$

Subtracting the two equations yields

$$0 = (\lambda_j - \lambda_k)\mathbf{u}_k^T \mathbf{u}_j$$

since $\mathbf{u}_k^T \mathbf{u}_j = \mathbf{u}_j^T \mathbf{u}_k$ and $\mathbf{u}_k^T B \mathbf{u}_j = \mathbf{u}_j^T B \mathbf{u}_k$. It follows that

$$\mathbf{u}_k^T \mathbf{u}_j \equiv (\mathbf{u}_k, \mathbf{u}_j) = 0$$

since $\lambda_j \neq \lambda_k$ by assumption.

Problem 5. Let

$$A = \begin{pmatrix} 0 & c & -b \\ -c & 0 & a \\ b & -a & 0 \end{pmatrix}$$

where $a, b, c \in \mathbb{R}$ and $a, b, c \neq 0$.
(i) Calculate the eigenvalues and eigenvectors of A.
(ii) Calculate $e^{\mu A}$ where $\mu \in \mathbb{R}$.

Solution 5.   (i) The matrix A is skew symmetric. Therefore the eigenvalues must be purely imaginary or zero. Since $\det A = 0$ and

$$\det A = \lambda_1 \lambda_2 \lambda_3$$

where λ_1, λ_2, λ_3 denote the eigenvalues we conclude that at least one of the eigenvalues must be zero. Thus the eigenvalues are 0, ki and $-ki$ where $k \in \mathbb{R}$. Eigenvalues of matrices over the field $\mathbb{R}$ occur in complex conjugate pairs. In the present case it also follows from the fact that

$$\mathrm{tr}A = \lambda_1 + \lambda_2 + \lambda_3 = 0.$$

To find k we have to solve $\det(A - \lambda I_3) = 0$. We obtain

$$\lambda^3 + \lambda(a^2 + b^2 + c^2) = 0.$$

Therefore

$$\lambda_1 = 0, \quad \lambda_2 = i\sqrt{a^2 + b^2 + c^2}, \quad \lambda_3 = -i\sqrt{a^2 + b^2 + c^2}.$$

The eigenvector of the eigenvalue $\lambda_1 = 0$ is determined by

$$A \begin{pmatrix} x_1 \\ x_2 \\ x_3 \end{pmatrix} = 0 \begin{pmatrix} x_1 \\ x_2 \\ x_3 \end{pmatrix} \equiv \begin{pmatrix} 0 \\ 0 \\ 0 \end{pmatrix}.$$

Consequently

$$cx_2 - bx_3 = 0, \quad -cx_1 + ax_3 = 0, \quad bx_1 - ax_2 = 0.$$

The solution is given by

$$\begin{pmatrix} x_1 \\ x_2 \\ x_3 \end{pmatrix} = \begin{pmatrix} a \\ b \\ c \end{pmatrix}.$$

The eigenvector of the eigenvalue $\lambda_2 = ik$ is determined by

$$cx_2 - bx_3 = ikx_1, \quad -cx_1 + ax_3 = ikx_2, \quad bx_1 - ax_2 = ikx_3$$

where

$$k := \sqrt{a^2 + b^2 + c^2}.$$

We find

$$\begin{pmatrix} x_1 \\ x_2 \\ x_3 \end{pmatrix} = \begin{pmatrix} ac - ibk \\ bc + iak \\ c^2 - k^2 \end{pmatrix}.$$

The eigenvector of the eigenvalue $\lambda_2 = -ik$ is determined by

$$cx_2 - bx_3 = -ikx_1, \quad -cx_1 + ax_3 = -ikx_2, \quad bx_1 - ax_2 = -ikx_3.$$

Solving this system of equations yields

$$\begin{pmatrix} x_1 \\ x_2 \\ x_3 \end{pmatrix} = \begin{pmatrix} ac + ibk \\ bc - iak \\ c^2 - k^2 \end{pmatrix}.$$

(ii) We calculate $e^{\mu A}$ in chapter 13, problem 2.

Problem 6. Let A be an $n \times n$ matrix. Show that

$$\det(\exp A) \equiv \exp(\mathrm{tr} A) \tag{1}$$

where tr(.) denotes the trace and det the determinant.

Solution 6. Any $n \times n$ matrix can be brought into a triangular form by a *similarity transformation*. This means there is an invertible $n \times n$ matrix R such that

$$R^{-1}AR = T \tag{2}$$

where T is a triangular matrix with diagonal elements t_{jj} which are the eigenvalues of A. We set $t_{jj} = \lambda_j$. From (2) it follows that

$$A = RTR^{-1}$$

and therefore

$$\exp(A) = \exp(RTR^{-1}) = R(\exp T)R^{-1}.$$

Since T is triangular, the diagonal elements of the k-th power of T are λ_j^k where k is a positive integer. Consequently, the diagonal elements of $\exp(T)$ are $\exp \lambda_j$. Since the determinant of a triangular matrix is equal to the product of its diagonal elements we find

$$\det(\exp T) = \exp(\lambda_1 + \lambda_2 + \cdots + \lambda_n) = \exp(\mathrm{tr} T).$$

Since

$$\mathrm{tr} T = \mathrm{tr}(R^{-1}AR) = \mathrm{tr} A$$

and

$$\det(\exp T) = \det(R(\exp T)R^{-1}) = \det(\exp(RTR^{-1})) = \det(\exp A)$$

we obtain (1). The identity $\exp(RTR^{-1}) \equiv R(\exp T)R^{-1}$ can easily be seen from

$$\exp(RTR^{-1}) = \sum_{k=0}^{\infty} \frac{(RTR^{-1})^k}{k!}$$

and $RR^{-1} = R^{-1}R = I_n$.

Problem 7.   Let A be a 4×4 symmetric matrix. Assume that the eigenvalues are given by 0, 1, 2, and 3 with the corresponding normalized eigenvectors

$$\frac{1}{\sqrt{2}}\begin{pmatrix} 1 \\ 0 \\ 0 \\ 1 \end{pmatrix}, \quad \frac{1}{\sqrt{2}}\begin{pmatrix} 1 \\ 0 \\ 0 \\ -1 \end{pmatrix}, \quad \frac{1}{\sqrt{2}}\begin{pmatrix} 0 \\ 1 \\ 1 \\ 0 \end{pmatrix}, \quad \frac{1}{\sqrt{2}}\begin{pmatrix} 0 \\ 1 \\ -1 \\ 0 \end{pmatrix}. \tag{1}$$

Find the matrix A.

Solution 7.   Since A is a symmetric matrix over $\mathbb{R}$ there exists an orthogonal matrix O such that

$$D = OAO^T \tag{2}$$

where D is the diagonal matrix

$$D = \text{diag}(0, 1, 2, 3).$$

The orthogonal matrix O^T is given by the normalized eigenvectors of A, i.e.

$$O^T = \frac{1}{\sqrt{2}}\begin{pmatrix} 1 & 1 & 0 & 0 \\ 0 & 0 & 1 & 1 \\ 0 & 0 & 1 & -1 \\ 1 & -1 & 0 & 0 \end{pmatrix}.$$

Thus

$$O = \frac{1}{\sqrt{2}}\begin{pmatrix} 1 & 0 & 0 & 1 \\ 1 & 0 & 0 & -1 \\ 0 & 1 & 1 & 0 \\ 0 & 1 & -1 & 0 \end{pmatrix}.$$

Since $O^T = O^{-1}$ we find from (2) that $A = O^T D O$. We obtain

$$A = \frac{1}{2}\begin{pmatrix} 1 & 0 & 0 & -1 \\ 0 & 5 & -1 & 0 \\ 0 & -1 & 5 & 0 \\ -1 & 0 & 0 & 1 \end{pmatrix}.$$

Note that we could also use the spectral theorem to find A.

Problem 8.   Let $\hat{B}$ and $\hat{C}$ be two linear bounded operators with a discrete spectrum (for example two finite-dimensional matrices). Assume that

$$[\hat{B}, \hat{C}]_+ = 0 \tag{1}$$

where the *anticommutator* is defined as

$$[\hat{B}, \hat{C}]_+ := \hat{B}\hat{C} + \hat{C}\hat{B}. \tag{2}$$

Let **u** be an eigenvector of both $\hat{B}$ and $\hat{C}$. What can be said about the corresponding eigenvalues?

Solution 8. From the eigenvalue equations

$$\hat{B}\mathbf{u} = B\mathbf{u}, \qquad \hat{C}\mathbf{u} = C\mathbf{u}$$

where B and C are the eigenvalues of $\hat{B}$ and $\hat{C}$, respectively, we obtain using (1)

$$[\hat{B}, \hat{C}]_+ \mathbf{u} = (\hat{B}\hat{C} + \hat{C}\hat{B})\mathbf{u} = \hat{B}(\hat{C}\mathbf{u}) + \hat{C}(\hat{B}\mathbf{u}).$$

Thus

$$[\hat{B}, \hat{C}]_+ \mathbf{u} = \hat{B}C\mathbf{u} + \hat{C}B\mathbf{u} = BC\mathbf{u} + CB\mathbf{u}.$$

Finally

$$[\hat{B}, \hat{C}]_+ \mathbf{u} = 2BC\mathbf{u} = \mathbf{0}.$$

Consequently, since $\mathbf{u} \neq \mathbf{0}$ we have the solutions

$$B = 0, \qquad C \neq 0$$

$$B \neq 0, \qquad C = 0$$

$$B = 0, \qquad C = 0.$$

Note that $[\,,\,]_+$ is called the *anticommutator.*

Problem 9. Let A be an $n \times n$ matrix with eigenvalue λ.
(i) Show that λ^2 is an eigenvalue of A^2.
(ii) Show that e^λ is an eigenvalue of e^A.
(iii) Show that $\sin(\lambda)$ is an eigenvalue of $\sin(A)$.
(iv) Assume that A^{-1} exists. Show that $1/\lambda$ is an eigenvalue of A^{-1}.

Solution 9. (i) From the eigenvalue equation $A\mathbf{x} = \lambda\mathbf{x}$ we obtain

$$A^2\mathbf{x} = A(A\mathbf{x}) = A(\lambda\mathbf{x}) = \lambda A\mathbf{x} = \lambda^2\mathbf{x}.$$

Obviously, $A^3\mathbf{x} = \lambda^3\mathbf{x}$ etc.
(ii) Using the expansion

$$e^A := \sum_{k=0}^{\infty} \frac{A^k}{k!}$$

and the result from (i) we find that e^A has the eigenvalue e^λ.

(iii) Using the expansion

$$\sin(A) := \sum_{k=0}^{\infty} \frac{(-1)^k A^{2k+1}}{(2k+1)!}$$

and the result from (i) we find that $\sin(\lambda)$ is an eigenvalue of $\sin(A)$.
(iv) From the eigenvalue equation (1) we find

$$A^{-1}(A\mathbf{x}) = A^{-1}(\lambda \mathbf{x}).$$

Thus $\mathbf{x} = \lambda A^{-1}\mathbf{x}$. Since $\mathbf{x} \neq \mathbf{0}$ we find that $1/\lambda$ is an eigenvalue of A^{-1}.

Problem 10.   A norm of an $n \times n$ matrix A over the real numbers can be defined as

$$\|A\| := \sup_{\|\mathbf{x}\|=1} \|A\mathbf{x}\| \tag{1}$$

where $\| \ \|$ on the right-hand side denotes the Euclidean norm. Let

$$A = \begin{pmatrix} 1 & 1 \\ 2 & 2 \end{pmatrix}.$$

(i) Find $\|A\|$.
(ii) Find the eigenvalues of $A^T A$ and compare with the result of (i).

Solution 10.   (i) We apply the method of the Lagrange multiplier. Let

$$\mathbf{x} := \begin{pmatrix} x_1 \\ x_2 \end{pmatrix}, \qquad \mathbf{x}^T = (\,x_1 \quad x_2\,)\,.$$

Using matrix multiplication we obtain

$$\|A\mathbf{x}\|^2 = (A\mathbf{x})^T A\mathbf{x} = \mathbf{x}^T A^T A\mathbf{x} = 5x_1^2 + 10x_1 x_2 + 5x_2^2\,.$$

The constraint is

$$\|\mathbf{x}\| = 1 \iff x_1^2 + x_2^2 = 1\,.$$

Thus we have

$$f(x_1, x_2) = 5x_1^2 + 10x_1 x_2 + 5x_2^2 + \lambda(x_1^2 + x_2^2 - 1)$$

where λ is the Lagrange multiplier. From the equations

$$\frac{\partial f}{\partial x_1} = 10x_1 + 10x_2 + 2\lambda x_1 = 0$$

$$\frac{\partial f}{\partial x_2} = 10x_1 + 10x_2 + 2\lambda x_2 = 0$$

we find $x_1^2 = x_2^2 = \dfrac{1}{2}$. Thus the square of the norm is given by

$$\|A\|^2 = 5 \cdot \frac{1}{2} + 10 \cdot \frac{1}{2} + 5 \cdot \frac{1}{2} = 10.$$

(ii) We have

$$A^T A = \begin{pmatrix} 5 & 5 \\ 5 & 5 \end{pmatrix}.$$

The rank of the matrix $A^T A$ is 1 and

$$\mathrm{tr}(A^T A) = 10.$$

Thus we find that the eigenvalues of $A^T A$ are given by 0 and 10. Thus the square of the norm is the largest eigenvalue of $A^T A$. Is this true in general?

Problem 11. Let

$$\mathbf{x} = \begin{pmatrix} x_1 \\ x_2 \\ x_3 \end{pmatrix}$$

where $x_j \in \mathbb{R}$. Show that at least one the eigenvalues of the 3×3 matrix

$$A := \begin{pmatrix} x_1 \\ x_2 \\ x_3 \end{pmatrix} \begin{pmatrix} x_1 & x_2 & x_3 \end{pmatrix}$$

is equal to zero.

Solution 11. We have

$$A = \begin{pmatrix} x_1^2 & x_1 x_2 & x_1 x_3 \\ x_1 x_2 & x_2^2 & x_2 x_3 \\ x_1 x_3 & x_2 x_3 & x_3^2 \end{pmatrix}.$$

Straightforward calculation yields $\det(A) = 0$. Since $\det(A) = \lambda_1 \lambda_2 \lambda_3$ we can conclude that at least one eigenvalue is 0. What are the other two eigenvalues?

Problem 12. Let A be an $n \times n$ matrix over $\mathbb{R}$. Assume that A^{-1} exists. To compute A^{-1} we can use the iteration

$$X_{t+1} = 2X_t - X_t A X_t, \qquad t = 0, 1, 2, \ldots$$

with $X_0 = \alpha A^T$. The scalar α is chosen such that $0 < \alpha < 2/\sigma_1$ with σ_1 to be the largest *singular value* of A. The singular values of a square matrix

B are the positive roots of the eigenvalues of the hermitian matrix B^*B (or B^TB if the matrix B is real). Apply the iteration to the 2×2 matrix

$$A = \begin{pmatrix} 1 & 1 \\ 1 & 0 \end{pmatrix}.$$

Solution 12.   The largest eigenvalue of A^TA is

$$\lambda = (3 + \sqrt{5})/2.$$

Thus we choose $\alpha = 1/2$ since

$$\frac{1}{2} < \frac{2\sqrt{2}}{\sqrt{3 + \sqrt{5}}}.$$

We obtain

$$X_1 = 2 \cdot \frac{1}{2} \begin{pmatrix} 1 & 1 \\ 1 & 0 \end{pmatrix} - \frac{1}{4} \begin{pmatrix} 1 & 1 \\ 1 & 0 \end{pmatrix} \begin{pmatrix} 1 & 1 \\ 1 & 0 \end{pmatrix} \begin{pmatrix} 1 & 1 \\ 1 & 0 \end{pmatrix} = \begin{pmatrix} 1/4 & 1/2 \\ 1/2 & -1/4 \end{pmatrix}.$$

For X_2 we find

$$X_2 = \begin{pmatrix} 3/16 & 11/16 \\ 11/16 & -1/2 \end{pmatrix}.$$

The series converges to the inverse of A

$$A^{-1} = \begin{pmatrix} 0 & 1 \\ 1 & -1 \end{pmatrix}.$$

Problem 13.   Let $\mathbf{x}$ be a nonzero column vector in $\mathbb{R}^n$ and $n \geq 2$. Consider the $n \times n$ matrix $\mathbf{x}\mathbf{x}^T$. Find one nonzero eigenvalue and the corresponding eigenvector of this matrix.

Solution 13.   Since matrix multiplication is associative we have

$$(\mathbf{x}\mathbf{x}^T)\mathbf{x} = \mathbf{x}(\mathbf{x}^T\mathbf{x}) = (\mathbf{x}^T\mathbf{x})\mathbf{x}.$$

Thus $\mathbf{x}$ is an eigenvector and $\mathbf{x}^T\mathbf{x}$ is the nonzero eigenvalue.

Problem 14.   (i) Let A, B be 2×2 matrices over $\mathbb{R}$ and vectors $\mathbf{x}$, $\mathbf{y}$ in $\mathbb{R}^2$ such that

$$A\mathbf{x} = \mathbf{y}, \qquad B\mathbf{y} = \mathbf{x}$$

and $\mathbf{x}^T\mathbf{y} = 0$, $\mathbf{x}^T\mathbf{x} = 1$, $\mathbf{y}^T\mathbf{y} = 1$. Show that AB and BA have the eigenvalue $+1$.

(ii) Find all 2×2 matrices A, B which satisfy the conditions given in (i). Use

$$\mathbf{x} = \begin{pmatrix} \cos \alpha \\ \sin \alpha \end{pmatrix}, \qquad \mathbf{y} = \begin{pmatrix} -\sin \alpha \\ \cos \alpha \end{pmatrix}.$$

Solution 14. (i) We have

$$B(A\mathbf{x}) = B\mathbf{y} = \mathbf{x} = (BA)\mathbf{x}$$
$$A(B\mathbf{y}) = A\mathbf{x} = \mathbf{y} = (AB)\mathbf{y}.$$

Thus the matrices AB and BA have the eigenvalue $+1$.
(ii) From

$$A\mathbf{x} = \begin{pmatrix} a_{11} & a_{12} \\ a_{21} & a_{22} \end{pmatrix} \begin{pmatrix} \cos \alpha \\ \sin \alpha \end{pmatrix} = \begin{pmatrix} -\sin \alpha \\ \cos \alpha \end{pmatrix} = \mathbf{y}$$

we obtain two equations for the unknown $a_{11}, a_{12}, a_{21}, a_{22}$

$$a_{11} \cos \alpha + (a_{12} + 1) \sin \alpha) = 0$$
$$(a_{21} - 1) \cos \alpha + a_{22} \sin \alpha = 0.$$

We eliminate a_{12} and a_{21}. The cases $\cos \alpha = 0$, $\sin \alpha \neq 0$ and $\sin \alpha = 0$, $\cos \alpha \neq 0$ are obvious. Assume now $\sin \alpha \neq 0$ and $\cos \alpha \neq 0$. Then we obtain the matrix

$$A = \begin{pmatrix} a_{11} & -1 - a_{11} \cos \alpha / \sin \alpha \\ 1 - a_{22} \sin \alpha / \cos \alpha & a_{22} \end{pmatrix}.$$

Similarly for the matrix B.

Problem 15. Find all 2×2 matrices over $\mathbb{R}$ such that the eigenvectors do not span $\mathbb{R}^2$.

Solution 15. Let

$$A = \begin{pmatrix} a & b \\ c & d \end{pmatrix}$$

be such a matrix. Thus A has at least one eigenvalue. If the eigenvectors of A do not span $\mathbb{R}^2$ then A has exactly one eigenvalue λ. Thus $\operatorname{tr} A = a + d = 2\lambda$ and $\det A = ad - bc = \lambda^2$ from which follows $(a + d)^2 = 4(ad - bc)$ which can be written as $(a - d)^2 = -bc$. Let $\mathbf{x} := (x_1 \quad x_2)^T$ be the corresponding eigenvector, i.e.

$$ax_1 + bx_2 = \lambda x_1 \quad \Rightarrow \quad ax_1 = \lambda x_1 - bx_2$$
$$cx_1 + dx_2 = \lambda x_2 \quad \Rightarrow \quad dx_2 = \lambda x_2 - cx_1.$$

Let $\mathbf{x}_\perp := (-x_2 \quad x_1)^T$, i.e. $\mathbf{x}^T\mathbf{x}_\perp = 0$. If the eigenvectors of A do not span $\mathbb{R}^2$ then $\mathbf{x}^T A\mathbf{x}_\perp \neq 0$. Substituting the equations above yields

$$\mathbf{x}^T A\mathbf{x}_\perp = bx_1^2 + (dx_2)x_1 - (ax_1)x_2 + cx_2^2 = (b-c)(x_1^2 + x_2^2) \neq 0.$$

Since $\mathbf{x} \neq \mathbf{0}$ we have $b - c \neq 0$. Thus we have the conditions

$$(a-d)^2 = -bc, \qquad b \neq c$$

which could also be written as

$$(\text{tr} A)^2 = 4\det A, \qquad A \neq A^T.$$

Problem 16.   Show that if A is an $n \times m$ matrix and if B is an $m \times n$ matrix, then $\lambda \neq 0$ is an eigenvalue of the $n \times n$ matrix AB if and only if λ is an eigenvalue of the $m \times m$ matrix BA. Show that if $m = n$ then the conclusion is true even for $\lambda = 0$.

Solution 16.   If $\lambda \neq 0$ is an eigenvalue of AB and if $\mathbf{v}$ is an eigenvector of AB, then $AB\mathbf{v} = \lambda\mathbf{v} \neq \mathbf{0}$ since $\lambda \neq 0$ and $\mathbf{v} \neq \mathbf{0}$. Thus $B\mathbf{v} \neq \mathbf{0}$. Therefore $BAB\mathbf{v} = \lambda B\mathbf{v}$ and λ is an eigenvalue of BA. If A and B are square matrices and $\lambda = 0$ is an eigenvalue of AB with eigenvector $\mathbf{v}$, then we have $\det(AB) = \det(BA) = 0$. Thus $\lambda = 0$ is an eigenvalue of BA.

Problem 17.   We know that a hermitian matrix has only real eigenvalues. Can we conclude that a matrix with only real eigenvalues is hermitian?

Solution 17.   We cannot conclude that a matrix with real eigenvalue is hermitian. For example the non-hermitian matrix

$$A = \begin{pmatrix} 0 & 1 & 0 \\ 1 & 0 & 1 \\ 0 & 2 & 0 \end{pmatrix}$$

admits the eigenvalues $\sqrt{3}$, $-\sqrt{3}$, 0.

Problem 18.   Let $a, b \in \mathbb{R}$. Find, by inspection, two eigenvectors and the corresponding eigenvalues of the 4×4 matrix

$$\begin{pmatrix} a & 0 & 0 & b \\ 0 & a & 0 & b \\ 0 & 0 & a & b \\ b & b & b & 0 \end{pmatrix}.$$

Solution 18.   Obviously

$$\begin{pmatrix} 1 \\ -1 \\ 0 \\ 0 \end{pmatrix}, \quad \begin{pmatrix} 1 \\ 0 \\ -1 \\ 0 \end{pmatrix}$$

are eigenvectors with the correponding eigenvalues a and a.

Problem 19.   Consider the following 3×3 matrix A and vector $\mathbf{v}$ in $\mathbb{R}^3$

$$A = \begin{pmatrix} 0 & 1 & 0 \\ 1 & 0 & 1 \\ 0 & 1 & 0 \end{pmatrix}, \quad \mathbf{v} = \begin{pmatrix} \sin(\alpha) \\ \sin(2\alpha) \\ \sin(3\alpha) \end{pmatrix}$$

where $\alpha \in \mathbb{R}$ and $\alpha \neq n\pi$ with $n \in \mathbb{Z}$. Show that using this vector we can find the eigenvalues and eigenvectors of A.

Solution 19.   We find

$$A\mathbf{v} = \begin{pmatrix} \sin(2\alpha) \\ \sin(\alpha) + \sin(3\alpha) \\ \sin(2\alpha) \end{pmatrix}.$$

Under the assumption that $\mathbf{v}$ is an eigenvector of A for specific α's we find the condition

$$\begin{pmatrix} \sin(2\alpha) \\ \sin(\alpha) + \sin(3\alpha) \\ \sin(2\alpha) \end{pmatrix} = \lambda \begin{pmatrix} \sin(\alpha) \\ \sin(2\alpha) \\ \sin(3\alpha) \end{pmatrix}.$$

Thus we have to solve the three equations

$$\sin(2\alpha) = \lambda \sin(\alpha)$$
$$\sin(\alpha) + \sin(3\alpha) = \lambda \sin(2\alpha)$$
$$\sin(2\alpha) = \lambda \sin(3\alpha).$$

We have to study the case $\lambda = 0$ and $\lambda \neq 0$. For $\lambda = 0$ we have to solve the two equations

$$\sin(2\alpha) = 0, \qquad \cos(2\alpha)\sin(\alpha) = 0$$

with the solution $\alpha = \pi/2$. Thus the eigenvalue $\lambda = 0$ has the normalized eigenvector

$$\frac{1}{\sqrt{2}} \begin{pmatrix} 1 \\ 0 \\ -1 \end{pmatrix}.$$

For $\lambda \neq 0$ we obtain the equation

$$2\sin(\alpha) = \lambda^2 \sin(\alpha)$$

after elimination of $\sin(2\alpha)$ and $\sin(3\alpha)$ using the first and third equation. Since $\alpha \neq n\pi$ we have that $\sin(\alpha) \neq 0$. Thus $\lambda^2 = 2$ with the eigenvalues $\lambda = -\sqrt{2}$ and $\lambda = \sqrt{2}$ and the corresponding normalized eigenvectors

$$\frac{1}{2}\begin{pmatrix} 1 \\ -\sqrt{2} \\ 1 \end{pmatrix}, \qquad \frac{1}{2}\begin{pmatrix} 1 \\ \sqrt{2} \\ 1 \end{pmatrix}.$$

Can this technqiue extended to the 4×4 matrix and vector $\mathbf{v} \in \mathbb{R}^4$

$$A = \begin{pmatrix} 0 & 1 & 0 & 0 \\ 1 & 0 & 1 & 0 \\ 0 & 1 & 0 & 1 \\ 0 & 0 & 1 & 0 \end{pmatrix}, \qquad \mathbf{v} = \begin{pmatrix} \sin(\alpha) \\ \sin(2\alpha) \\ \sin(3\alpha) \\ \sin(4\alpha) \end{pmatrix}$$

and even higher dimensions?

Problem 20.   Let B be a 2×2 matrix with eigenvalues λ_1 and λ_2. Find the eigenvalues of the 4×4 matrix

$$X = \begin{pmatrix} 0 & 0 & 1 & 0 \\ 0 & 0 & 0 & 1 \\ b_{11} & b_{12} & 0 & 0 \\ b_{21} & b_{22} & 0 & 0 \end{pmatrix}$$

Let $\mathbf{v}$ be an eigenvector of B with eigenvalue λ. What can be said about an eigenvector of the 4×4 matrix X given by eigenvector $\mathbf{v}$ and eigenvalue of B.

Solution 20.   The characteristic equation $\det(X - rI_4) = 0$ is given by

$$r^4 - r^2(b_{11} + b_{22}) + (b_{11}b_{22} - b_{12}b_{21}) = 0$$

or

$$r^4 - r^2 \mathrm{tr}(B) + \det(B) = 0.$$

Thus since $\mathrm{tr}(B) = \lambda_1 + \lambda_2$ and $\det(B) = \lambda_1\lambda_2$ we have

$$r^4 - r^2(\lambda_1 + \lambda_2) + \lambda_1\lambda_2 = 0.$$

This quartic equation can be reduced to a quadratic equation. We find the eigenvalues

$$r_1 = \sqrt{\lambda_1}, \quad r_2 = -\sqrt{\lambda_1}, \quad r_3 = \sqrt{\lambda_2}, \quad r_4 = -\sqrt{\lambda_2}.$$

Let **v** be an eigenvector of B with eigenvalue λ, i.e. $B\mathbf{v} = \lambda\mathbf{v}$. Then we have

$$
\begin{pmatrix} 0 & 0 & 1 & 0 \\ 0 & 0 & 0 & 1 \\ b_{11} & b_{12} & 0 & 0 \\ b_{21} & b_{22} & 0 & 0 \end{pmatrix} \begin{pmatrix} v_1 \\ v_2 \\ \sqrt{\lambda}v_1 \\ \sqrt{\lambda}v_2 \end{pmatrix} = \sqrt{\lambda} \begin{pmatrix} v_1 \\ v_2 \\ \sqrt{\lambda}v_1 \\ \sqrt{\lambda}v_2 \end{pmatrix}.
$$

Thus

$$
\begin{pmatrix} v_1 \\ v_2 \\ \sqrt{\lambda}v_1 \\ \sqrt{\lambda}v_2 \end{pmatrix}
$$

is an eigenvector of X.

Problem 21.   Consider the infinite-dimensional symmetric matrix

$$
A = \begin{pmatrix} 0 & 1 & 0 & 0 & \cdots \\ 1 & 0 & 1 & 0 & \cdots \\ 0 & 1 & 0 & 1 & \cdots \\ & & \ddots & & \ddots \\ & & & \ddots & & \ddots \\ & & & & \ddots \end{pmatrix}.
$$

In other words

$$
A_{jk} = \begin{cases} 1 & \text{if} \quad j = k+1 \\ 1 & \text{if} \quad j = k-1 \\ 0 & \text{otherwise} \end{cases}
$$

with $j, k = 1, 2, 3, \ldots$.   Find the spectrum of this infinite-dimensional matrix.

Solution 21.   Let A_n be the $n \times n$ truncated matrix of A. Then the eigenvalue problem for A_n is given by $A_n\mathbf{x} = \lambda\mathbf{x}$ with

$$
A_n = \begin{pmatrix} 0 & 1 & 0 & 0 & \cdots & 0 & 0 & 0 \\ 1 & 0 & 1 & 0 & \cdots & 0 & 0 & 0 \\ 0 & 1 & 0 & 1 & \cdots & 0 & 0 & 0 \\ & & \ddots & & \ddots \\ & \vdots & & \ddots & & \ddots \\ & & & & \ddots & & \ddots \\ 0 & 0 & 0 & 0 & \cdots & 1 & 0 & 1 \\ 0 & 0 & 0 & 0 & \cdots & 0 & 1 & 0 \end{pmatrix}.
$$

First we calculate the eigenvalues of A_n. Then we study A_n as $n \to \infty$.

The eigenvalue problem leads to $D_n(\lambda) = 0$ where

$$D_n(\lambda) \equiv \det \begin{pmatrix} -\lambda & 1 & 0 & 0 & \cdots & 0 & 0 & 0 \\ 1 & -\lambda & 1 & 0 & \cdots & 0 & 0 & 0 \\ 0 & 1 & -\lambda & 1 & \cdots & 0 & 0 & 0 \\ & & \vdots & & & & & \\ 0 & 0 & 0 & 0 & \cdots & 1 & -\lambda & 1 \\ 0 & 0 & 0 & 0 & \cdots & 0 & 1 & -\lambda \end{pmatrix}.$$

We try to find a difference equation for $D_n(\lambda)$, where $n = 1, 2, \ldots$. We obtain

$$D_n(\lambda) = -\lambda \det \begin{pmatrix} -\lambda & 1 & 0 & 0 & \cdots & 0 & 0 & 0 \\ 1 & -\lambda & 1 & 0 & \cdots & 0 & 0 & 0 \\ 0 & 1 & -\lambda & 1 & \cdots & 0 & 0 & 0 \\ 0 & 0 & 1 & -\lambda & \cdots & 0 & 0 & 0 \\ & & \vdots & & & & & \\ 0 & 0 & 0 & 0 & \cdots & 1 & -\lambda & 1 \\ 0 & 0 & 0 & 0 & \cdots & 0 & 1 & -\lambda \end{pmatrix}$$

$$- \det \begin{pmatrix} 1 & 1 & 0 & 0 & \cdots & 0 & 0 & 0 \\ 0 & -\lambda & 1 & 0 & \cdots & 0 & 0 & 0 \\ 0 & 1 & -\lambda & 1 & \cdots & 0 & 0 & 0 \\ & & \vdots & & & & & \\ 0 & 0 & 0 & 0 & \cdots & 1 & -\lambda & 1 \\ 0 & 0 & 0 & 0 & \cdots & 0 & 1 & -\lambda \end{pmatrix}.$$

The first determinant on the right-hand side is equal to $D_{n-1}(\lambda)$. For the second determinant we find (expansion of the first row)

$$\det \begin{pmatrix} 1 & 1 & 0 & 0 & \cdots & 0 & 0 & 0 \\ 0 & -\lambda & 1 & 0 & \cdots & 0 & 0 & 0 \\ 0 & 1 & -\lambda & 1 & \cdots & 0 & 0 & 0 \\ & & \vdots & & & & & \\ 0 & 0 & 0 & 0 & \cdots & 1 & -\lambda & 1 \\ 0 & 0 & 0 & 0 & \cdots & 0 & 1 & -\lambda \end{pmatrix} = D_{n-2}(\lambda).$$

Consequently, we obtain a second order linear difference equation

$$D_n(\lambda) = -\lambda D_{n-1}(\lambda) - D_{n-2}(\lambda)$$

with the "initial condition"

$$D_1(\lambda) = -\lambda, \qquad D_2(\lambda) = \lambda^2 - 1.$$

To solve this linear difference equation we make the ansatz

$$D_n(\lambda) = e^{in\theta}$$

where $n = 1, 2, \ldots$. Inserting the this ansatz into the difference equation yields

$$e^{in\theta} = -\lambda e^{i(n-1)\theta} - e^{i(n-2)\theta}.$$

It follows that $e^{i\theta} = -\lambda - e^{-i\theta}$. Consequently $\lambda = -2\cos\theta$. The general solution to the linear difference equation is given by

$$D_n(\lambda) = C_1 \cos(n\theta) + C_2 \sin(n\theta)$$

where C_1 and C_2 are constants and $\lambda = -2\cos\theta$. Imposing the initial conditions, it follows that

$$D_n(\lambda) = \frac{\sin(n+1)\theta}{\sin\theta}.$$

Since $D_n(\lambda) = 0$ we have to solve the equation

$$\frac{\sin(n+1)\theta}{\sin\theta} = 0.$$

The solutions to this equation are given by $\theta = k\pi/(n+1)$ with $k = 1, 2, \ldots, n$. Since $\lambda = -2\cos\theta$, we find the eigenvalues

$$\lambda_k = -2\cos\left(\frac{k\pi}{n+1}\right)$$

with $k = 1, 2, \ldots, n$. Consequently, $|\lambda_k| \leq 2$. If $n \to \infty$, then there are infinitely many λ_k with $|\lambda_k| \leq 2$ and $\lambda_k - \lambda_{k+1} \to 0$. Therefore

$$\text{spectrum} A = [-2, 2]$$

i.e., we have a *continuous spectrum*.

Another approach to find the spectrum is as follows. Let

$$A = B + B^T$$

where

$$B = \begin{pmatrix} 0 & 0 & 0 & 0 & \cdots \\ 1 & 0 & 0 & 0 & \cdots \\ 0 & 1 & 0 & 0 & \cdots \\ 0 & 0 & 1 & 0 & \cdots \\ \vdots & & & & \end{pmatrix}, \qquad B^T = \begin{pmatrix} 0 & 1 & 0 & 0 & \cdots \\ 0 & 0 & 1 & 0 & \cdots \\ 0 & 0 & 0 & 1 & \cdots \\ \vdots & & & & \end{pmatrix}.$$

Then

$$B^T B = I$$

where I is the infinite unit matrix. Notice that $BB^T \neq I$. We use the following notation

$$Cf = \lambda f \quad \text{means} \quad \|Cf_n - \lambda f_n\| \to 0.$$

Now
$$Bf = \lambda f \quad \Rightarrow \quad B^T B f = B^T \lambda f.$$

Therefore
$$\Rightarrow \quad If = \lambda B^T f \quad \Rightarrow \quad f = \lambda B^T f \quad \Rightarrow \quad B^T f = \frac{1}{\lambda} f.$$

From $Bf = \lambda f$ it also follows that
$$\|Bf\|^2 = (Bf, Bf) = (f, B^T B f) = (f, f) = \|f\|^2.$$

On the other hand
$$(Bf, Bf) = \bar{\lambda}\lambda(f, f) = |\lambda|^2 \|f\|^2.$$

Since $\|f\| > 0$ we find that $|\lambda|^2 = 1$. Therefore
$$Af = (B + B^T)f = \left(\lambda + \frac{1}{\lambda}\right) f = (\lambda + \bar{\lambda})f = 2(\cos\phi)f.$$

This means
$$\bigwedge_{\phi \in \mathbf{R}} \|(A - 2(\cos\phi)I)f_n\| \to 0$$

or
$$A \begin{pmatrix} \sin\phi \\ \sin(2\phi) \\ \sin(3\phi) \\ \vdots \end{pmatrix} = 2\cos\phi \begin{pmatrix} \sin\phi \\ \sin(2\phi) \\ \sin(3\phi) \\ \vdots \end{pmatrix}.$$

The vector on the left-hand side is not an element of the Hilbert space $\ell_2(\mathbb{N})$. For the first two rows we have the identities
$$\sin(2\phi) = 2\sin\phi\cos\phi, \qquad \sin\phi + \sin(3\phi) = 2\cos\phi\sin(2\phi).$$

Here $\ell_2(\mathbb{N})$ is the *Hilbert space* of all infinite vectors (sequences)
$$\mathbf{u} = (u_1, u_2, \ldots)^T$$
of complex numbers u_j such that
$$\sum_{j=1}^{\infty} |u_j|^2 < \infty.$$

Chapter 8

Functions of Matrices

Problem 1. Let A and L be two $n \times n$ matrices. Calculate

$$e^L A e^{-L}.$$

Solution 1. We consider

$$A(\epsilon) := e^{\epsilon L} A e^{-\epsilon L}$$

where ϵ is a real parameter. Taking the derivative of $A(\epsilon)$ with respect to ϵ yields

$$\frac{dA}{d\epsilon} = L e^{\epsilon L} A e^{-\epsilon L} - e^{\epsilon L} A e^{-\epsilon L} L = [L, A(\epsilon)].$$

The second derivative gives

$$\frac{d^2 A}{d\epsilon^2} = \left[L, \frac{dA}{d\epsilon} \right] = [L, [L, A(\epsilon)]]$$

and so on. Consequently, we can write the matrix $e^L A e^{-L} = A(1)$ as a *Taylor series expansion* about the origin, i.e.,

$$A(1) = A(0) + \frac{1}{1!} \frac{dA(0)}{d\epsilon} + \frac{1}{2!} \frac{d^2 A(0)}{d\epsilon^2} + \cdots$$

where $A(0) = A$ and

$$\frac{dA(0)}{d\epsilon} \equiv \frac{dA(\epsilon)}{d\epsilon} \bigg|_{\epsilon=0}.$$

Thus we find

$$e^L A e^{-L} \equiv A + [L, A] + \frac{1}{2!}[L, [L, A]] + \frac{1}{3!}[L, [L, [L, A]]] + \cdots .$$

Problem 2.   Let A and B be two $n \times n$ matrices. Assume that

$$[[A, B], A] = 0, \qquad [[A, B], B] = 0. \tag{1}$$

Show that

$$\exp(A) \exp(B) \equiv \exp(A + B + \frac{1}{2}[A, B]). \tag{2}$$

Solution 2.   Let

$$T(\epsilon) := e^{\epsilon A} e^{\epsilon B} \tag{3}$$

where ϵ is a real parameter. Differentiating (3) with respect to ϵ gives

$$\frac{dT}{d\epsilon} = A e^{\epsilon A} e^{\epsilon B} + e^{\epsilon A} B e^{\epsilon B} = (A + e^{\epsilon A} B e^{-\epsilon A}) T(\epsilon). \tag{4}$$

Using (1), we find

$$e^{\epsilon A} B e^{-\epsilon A} = B - [B, A]\epsilon$$

Therefore

$$\frac{dT}{d\epsilon} = (A + B + [A, B]\epsilon) T(\epsilon).$$

Consequently, T is the solution to this matrix differential equation with the initial condition

$$T(\epsilon = 0) = I_n$$

where I_n is the $n \times n$ unit matrix. The initial condition follows from (3). Since the matrices $A + B$ and $[A, B]$ commute, the matrix differential equation (4) can be integrated as if the matrices were merely numbers, to give the solution

$$T(\epsilon) = e^{\epsilon(A+B)} e^{\frac{1}{2}\epsilon^2 [A,B]}.$$

The identity (2) follows by setting $\epsilon = 1$.

Problem 3.   Let $\boldsymbol{\sigma} := (\sigma_1, \sigma_2, \sigma_3)$, where

$$\sigma_1 := \begin{pmatrix} 0 & 1 \\ 1 & 0 \end{pmatrix}, \qquad \sigma_2 := \begin{pmatrix} 0 & -i \\ i & 0 \end{pmatrix}, \qquad \sigma_3 := \begin{pmatrix} 1 & 0 \\ 0 & -1 \end{pmatrix}$$

are the *Pauli spin matrices*. Let $\mathbf{a} := (a_1, a_2, a_3)$ where $a_j \in \mathbb{R}$ and

$$\mathbf{a} \cdot \boldsymbol{\sigma} := a_1 \sigma_1 + a_2 \sigma_2 + a_3 \sigma_3.$$

(i) Calculate $(\mathbf{a} \cdot \boldsymbol{\sigma})^2$, $(\mathbf{a} \cdot \boldsymbol{\sigma})^3$ and $(\mathbf{a} \cdot \boldsymbol{\sigma})^4$.
(ii) Calculate

$$\exp(\mathbf{a} \cdot \boldsymbol{\sigma}).$$

Solution 3. (i) First we notice that

$$\sigma_1^2 = I_2, \qquad \sigma_2^2 = I_2, \qquad \sigma_3^2 = I_2$$

where I_2 is the 2×2 unit matrix. Furthermore the anticommutator vanishes

$$\sigma_1 \sigma_2 + \sigma_2 \sigma_1 = 0, \qquad \sigma_1 \sigma_3 + \sigma_3 \sigma_1 = 0, \qquad \sigma_2 \sigma_3 + \sigma_3 \sigma_2 = 0$$

where 0 is the 2×2 zero matrix. Thus

$$(\mathbf{a} \cdot \boldsymbol{\sigma})^2 = (a_1 \sigma_1 + a_2 \sigma_2 + a_3 \sigma_3)(a_1 \sigma_1 + a_2 \sigma_2 + a_3 \sigma_3) = (a_1^2 + a_2^2 + a_3^2)I_2.$$

In the following we set

$$a^2 \equiv \mathbf{a}^2 := a_1^2 + a_2^2 + a_3^2.$$

It follows that

$$(\mathbf{a} \cdot \boldsymbol{\sigma})^3 = (\mathbf{a} \cdot \boldsymbol{\sigma})^2 (\mathbf{a} \cdot \boldsymbol{\sigma}) = (a^2 I_2)(\mathbf{a} \cdot \boldsymbol{\sigma}) = a^2 (\mathbf{a} \cdot \boldsymbol{\sigma})$$

and

$$(\mathbf{a} \cdot \boldsymbol{\sigma})^4 = (\mathbf{a} \cdot \boldsymbol{\sigma})^2 (\mathbf{a} \cdot \boldsymbol{\sigma})^2 = (a^2 I_2)(a^2 I) = a^4 I_2.$$

(ii) From the definition

$$\exp(\mathbf{a} \cdot \boldsymbol{\sigma}) := \sum_{k=0}^{\infty} \frac{(\mathbf{a} \cdot \boldsymbol{\sigma})^k}{k!}$$

and the result from (i) we have

$$\exp(\mathbf{a} \cdot \boldsymbol{\sigma}) = I_2 + \mathbf{a} \cdot \boldsymbol{\sigma} + \frac{(\mathbf{a} \cdot \boldsymbol{\sigma})^2}{2!} + \frac{(\mathbf{a} \cdot \boldsymbol{\sigma})^3}{3!} + \frac{(\mathbf{a} \cdot \boldsymbol{\sigma})^4}{4!} + \frac{(\mathbf{a} \cdot \boldsymbol{\sigma})^5}{5!} + \cdots$$

$$= I_2 + \mathbf{a} \cdot \boldsymbol{\sigma} + \frac{a^2 I_2}{2!} + \frac{a^2 (\mathbf{a} \cdot \boldsymbol{\sigma})}{3!} + \frac{a^4 I_2}{4!} + \frac{a^4 (\mathbf{a} \cdot \boldsymbol{\sigma})}{5!} + \cdots$$

$$= \left(1 + \frac{a^2}{2!} + \frac{a^4}{4!} + \cdots\right) I_2 + \left(1 + \frac{a^2}{3!} + \frac{a^4}{5!} + \cdots\right) \mathbf{a} \cdot \boldsymbol{\sigma}.$$

Thus

$$\exp(\mathbf{a} \cdot \boldsymbol{\sigma}) = (\cosh a)I_2 + \frac{1}{a}(\sinh a)(\mathbf{a} \cdot \boldsymbol{\sigma})$$

for $a \neq 0$. If $\mathbf{a} = (0,0,0)$, we have

$$\exp(\mathbf{a} \cdot \boldsymbol{\sigma}) = \begin{pmatrix} 1 & 0 \\ 0 & 1 \end{pmatrix}.$$

Problem 4.   Let

$$A = \begin{pmatrix} -1 & 2 \\ 2 & -1 \end{pmatrix}.$$ (1)

Calculate $\exp(\epsilon A)$ where $\epsilon \in \mathbb{R}$.

Solution 4.   The matrix A is symmetric. Therefore, there exists an orthogonal matrix U such that $U A U^{-1}$ is a diagonal matrix. The diagonal elements of $U A U^{-1}$ are the eigenvalues of A. The matrix U^* we find from the normalized eigenvectors of A. Since A is symmetric the eigenvalues are real. We set

$$D = U A U^{-1}$$ (2)

with

$$D = \begin{pmatrix} d_{11} & 0 \\ 0 & d_{22} \end{pmatrix}.$$

Then

$$\exp D = \begin{pmatrix} e^{d_{11}} & 0 \\ 0 & e^{d_{22}} \end{pmatrix}.$$

From (2) it follows that

$$e^{\epsilon D} = \exp(\epsilon U A U^{-1}) = U \exp(\epsilon A) U^{-1}.$$

Therefore

$$\exp(\epsilon A) = U^{-1} e^{\epsilon D} U.$$

The matrix U is constructed by means of the eigenvalues and normalized eigenvectors of A. The eigenvalues of A are given by

$$\lambda_1 = 1, \qquad \lambda_2 = -3.$$

The corresponding normalized eigenvectors are

$$\frac{1}{\sqrt{2}} \begin{pmatrix} 1 \\ 1 \end{pmatrix}, \qquad \frac{1}{\sqrt{2}} \begin{pmatrix} 1 \\ -1 \end{pmatrix}.$$

Consequently, the matrix U^* is given by

$$U^* = \frac{1}{\sqrt{2}} \begin{pmatrix} 1 & 1 \\ 1 & -1 \end{pmatrix}.$$

It follows that $U^* = U = U^{-1}$. Finally we arrive at

$$\exp(\epsilon A) = U^* e^{\epsilon D} U = \frac{1}{2} \begin{pmatrix} e^\epsilon + e^{-3\epsilon} & e^\epsilon - e^{-3\epsilon} \\ e^\epsilon - e^{-3\epsilon} & e^\epsilon + e^{-3\epsilon} \end{pmatrix}.$$

The solution of the initial value problem of the autonomous system of linear ordinary differential equations

$$\frac{du_1}{d\epsilon} = -u_1 + 2u_2$$

$$\frac{du_2}{d\epsilon} = 2u_1 - u_2$$

is given by

$$\begin{pmatrix} u_1(\epsilon) \\ u_2(\epsilon) \end{pmatrix} = e^{\epsilon A} \begin{pmatrix} u_1(\epsilon = 0) \\ u_2(\epsilon = 0) \end{pmatrix}.$$

Problem 5.   Let

$$\hat{H} = \hbar\omega\sigma_z$$

be a Hamilton operator acting in the two-dimensional Hilbert space $\mathbb{C}^2$, where

$$\sigma_z := \begin{pmatrix} 1 & 0 \\ 0 & -1 \end{pmatrix}$$

and ω is the frequency. Calculate the time evolution of

$$\sigma_x := \begin{pmatrix} 0 & 1 \\ 1 & 0 \end{pmatrix}$$

The matrices σ_x, σ_y and σ_z are the Pauli matrices, where

$$\sigma_y := \begin{pmatrix} 0 & -i \\ i & 0 \end{pmatrix}.$$

The Pauli matrices form a Lie algebra under the commutator. The *Heisenberg equation of motion* is given by

$$i\hbar\frac{d\sigma_x}{dt} = [\sigma_x, \hat{H}](t).$$

Solution 5.   Since

$$[\sigma_x, \hat{H}] = \hbar\omega[\sigma_x, \sigma_z] = -2i\hbar\omega\sigma_y$$

we obtain

$$\frac{d\sigma_x}{dt} = -2\omega\sigma_y(t).$$

Now we have to calculate the time-evolution of σ_y, i.e.,

$$i\hbar\frac{d\sigma_y}{dt} = [\sigma_y, \hat{H}](t).$$

Since
$$[\sigma_y, \hat{H}] = \hbar\omega[\sigma_y, \sigma_z] = 2i\hbar\omega\sigma_x$$

we find
$$\frac{d\sigma_y}{dt} = 2\omega\sigma_x(t).$$

Thus we have to solve the following system of linear matrix differential equations with constant coefficients

$$\frac{d\sigma_x}{dt} = -2\omega\sigma_y(t), \qquad \frac{d\sigma_y}{dt} = 2\omega\sigma_x(t)$$

together with the initial conditions

$$\sigma_x(t = 0) = \sigma_x, \qquad \sigma_y(t = 0) = \sigma_y.$$

The solution of the initial value problem is given by

$$\sigma_x(t) = \sigma_x \cos(2\omega t) - \sigma_y \sin(2\omega t)$$

$$\sigma_y(t) = \sigma_y \cos(2\omega t) + \sigma_x \sin(2\omega t).$$

The solution of the Heisenberg equation of motion can also be given as

$$\sigma_x(t) = e^{i\hat{H}t/\hbar}\sigma_x e^{-i\hat{H}t/\hbar}$$

$$\sigma_y(t) = e^{i\hat{H}t/\hbar}\sigma_y e^{-i\hat{H}t/\hbar}.$$

Problem 6.   Let $f : \mathbb{R} \to \mathbb{R}$ be an analytic function. Let $\theta \in \mathbb{R}$, $\mathbf{n}$ be a normalized vector in $\mathbb{R}^3$ and σ_1, σ_2, σ_3 be the Pauli spin matrices. We define
$$\mathbf{n} \cdot \boldsymbol{\sigma} := n_1\sigma_1 + n_2\sigma_2 + n_3\sigma_3.$$

Then

$$f(\theta\mathbf{n} \cdot \boldsymbol{\sigma}) \equiv \frac{1}{2}(f(\theta) + f(-\theta))I_2 + \frac{1}{2}(f(\theta) - f(-\theta))(\mathbf{n} \cdot \boldsymbol{\sigma}).$$

Apply this identity to $f(x) = \sin(x)$.

Solution 6.   Since $f(-x) = \sin(-x) = -\sin(x)$ we have $f(x) + f(-x) = 0$ and $f(x) - f(-x) = 2\sin(x)$. Thus we obtain

$$f(\theta\mathbf{n} \cdot \boldsymbol{\sigma}) = \sin(\theta)\mathbf{n} \cdot \boldsymbol{\sigma}.$$

Problem 7.   Let $\epsilon \in \mathbb{R}$. Calculate

$$f(\epsilon) = e^{-\epsilon\sigma_y}\sigma_z e^{\epsilon\sigma_y}.$$

Hint. Differentiate the matrix-valued function f with respect to ϵ and solve the initial value problem of the resulting ordinary differential equation.

Solution 7.   Since $[\sigma_y,\sigma_z]=2i\sigma_x$, $[\sigma_y,\sigma_x]=-2i\sigma_z$ we obtain

$$\frac{df}{d\epsilon}=-e^{-\epsilon\sigma_y}\sigma_y\sigma_z e^{\epsilon\sigma_y}+e^{-\epsilon\sigma_y}\sigma_z\sigma_y e^{\epsilon\sigma_y}$$
$$=-e^{-\epsilon\sigma_y}[\sigma_y,\sigma_z]e^{\epsilon\sigma_y}$$
$$=-2ie^{-\epsilon\sigma_y}\sigma_x e^{\epsilon\sigma_y}$$

and

$$\frac{d^2f}{d\epsilon^2}=2i(e^{-\epsilon\sigma_y}\sigma_y\sigma_x e^{\epsilon\sigma_y}-e^{-\epsilon\sigma_y}\sigma_x\sigma_y e^{\epsilon\sigma_y})$$
$$=-2ie^{-\epsilon\sigma_y}2i\sigma_z e^{\epsilon\sigma_y}$$
$$=4f(\epsilon)\,.$$

Thus we have to solve the linear differential equation with constant coefficients

$$\frac{d^2f}{d\epsilon^2}=4f\,.$$

The inital conditions are $f(0)=\sigma_z$ and $df(0)/d\epsilon=-2i\sigma_x$. We obtain the solution

$$f(\epsilon)=\frac{1}{2}(\sigma_z-i\sigma_x)e^{2\epsilon}+\frac{1}{2}(\sigma_z+i\sigma_x)e^{-2\epsilon}\,.$$

Problem 8.   Let A, B be $n\times n$ matrices over $\mathbb{C}$ and $\alpha\in\mathbb{C}$. The *Baker-Campbell-Hausdorff formula* states that

$$e^{\alpha A}Be^{-\alpha A}=B+\alpha[A,B]+\frac{\alpha^2}{2!}[A,[A,B]]+\cdots=\sum_{j=0}^{\infty}\frac{\alpha^j}{j!}\{A^j,B\}=\tilde{B}(\alpha)$$

where $[A,B]:=AB-BA$ and

$$\{A^j,B\}=[A,\{A^{j-1},B\}]$$

is the repeated commutator.
(i) Extend the formula to

$$e^{\alpha A}B^k e^{-\alpha A}$$

where $k>1$.
(ii) Extend the formula to

$$e^{\alpha A}e^B e^{-\alpha A}\,.$$

Solution 8.   (i) Using that $e^{-\alpha A}e^{\alpha A} = I_n$ we have

$$e^{\alpha A}B^k e^{-\alpha A} = e^{\alpha A}Be^{-\alpha A}e^{\alpha A}B^{k-1}e^{-\alpha A}$$
$$= \widetilde{B}(\alpha)e^{\alpha A}B^{k-1}e^{-\alpha A}$$
$$= (B(\alpha))^k .$$

(ii) Using the result from (i) we obtain

$$e^{\alpha A}e^{B}e^{-\alpha A} = \exp(\widetilde{B}(\alpha)) .$$

Problem 9.   Consider the matrix

$$A = \begin{pmatrix} \lambda_1 & c \\ 0 & \lambda_2 \end{pmatrix}$$

where $\lambda_1, \lambda_2, c \in \mathbb{R}$. Calculate $\sin(A)$.

Solution 9.   For the case $\lambda_1 = \lambda_2 = \lambda$ we obviously find

$$\sin\begin{pmatrix} \lambda & c \\ 0 & \lambda \end{pmatrix} = \begin{pmatrix} \sin\lambda & c\cos\lambda \\ 0 & \sin\lambda \end{pmatrix} .$$

Consider now the case $\lambda_1 \neq \lambda_2$. First we note that

$$\begin{pmatrix} \lambda_1 & c \\ 0 & \lambda_2 \end{pmatrix}^n = \begin{pmatrix} \lambda_1^n & s \\ 0 & \lambda_2^n \end{pmatrix}$$

where

$$s = (\lambda_1^{n-1} + \lambda_1^{n-2}\lambda_2 + \lambda_1^{n-3}\lambda_2^2 + \cdots + \lambda_2^{n-1})c .$$

Thus

$$\frac{s(\lambda_1 - \lambda_2)}{\lambda_1 - \lambda_2} = \frac{1}{\lambda_1 - \lambda_2}(\lambda_1^n - \lambda_2^n)c .$$

Thus

$$\sin(A) = \begin{pmatrix} \sin(\lambda_1) & c(\sin(\lambda_1) - \sin(\lambda_2))/(\lambda_1 - \lambda_2) \\ 0 & \sin(\lambda_2) \end{pmatrix} .$$

Chapter 9

Transformations

Problem 1. Let $f : [0,1] \mapsto [0,1]$ be defined by

$$f(x) = 4x(1-x) \tag{1}$$

and let $\phi : [0,1] \mapsto [0,1]$ be defined by

$$\phi(x) = \frac{2}{\pi} \arcsin \sqrt{x}.$$

Calculate

$$\widetilde{f} = \phi \circ f \circ \phi^{-1}$$

where $\circ$ denotes the *composition of functions*.

Solution 1. The inverse of ϕ is defined on $[0,1]$ and we find

$$\phi^{-1}(x) = \sin^2\left(\frac{\pi x}{2}\right) \equiv \frac{1 - \cos(\pi x)}{2}$$

since

$$\arcsin \circ \sin y = y, \qquad \sin \circ \arcsin x = x$$

where

$$y \in \left[-\frac{\pi}{2}, \frac{\pi}{2}\right].$$

To find $\widetilde{f}$ we set

$$y(x) = \sin^2\left(\frac{\pi x}{2}\right), \qquad v(y) = 4y(1-y), \qquad w(v) = \frac{2}{\pi} \arcsin \sqrt{v}.$$

Inserting $y(x)$ into $v(y)$ gives

$$v(x) = 4\sin^2\left(\frac{\pi x}{2}\right)\left(1 - \sin^2\frac{\pi x}{2}\right)$$
$$= 4\sin^2\left(\frac{\pi x}{2}\right)\cos^2\left(\frac{\pi x}{2}\right)$$
$$= \left(2\sin\left(\frac{\pi x}{2}\right)\cos\left(\frac{\pi x}{2}\right)\right)^2.$$

Now

$$\sqrt{v(x)} = 2\sin\left(\frac{\pi x}{2}\right)\cos\left(\frac{\pi x}{2}\right) = \sin\pi x.$$

For $x \in \left[0, \frac{1}{2}\right]$ we have $\pi x \in \left[0, \frac{\pi}{2}\right]$ and therefore we find for this range

$$w(x) = \frac{2}{\pi}\arcsin(\sin\pi x) = 2x.$$

For $x \in \left[\frac{1}{2}, 1\right]$ we have $\pi x \in \left[\frac{1}{2}\pi, \pi\right]$ and therefore we find for this range

$$w(x) = \frac{2}{\pi}\arcsin(\sin\pi x) = 2(1 - x).$$

Consequently,

$$\widetilde{f}(x) = \begin{cases} 2x & \text{if } x \in \left[0, \frac{1}{2}\right] \\ 2(1-x) & \text{if } x \in \left[\frac{1}{2}, 1\right]. \end{cases}$$

The mapping (1) is called the *logistic map* and mapping $\widetilde{f}$ is called the *tent map*. Both play an important role in the study of nonlinear systems with *chaotic behaviour*.

Problem 2.   (i) Consider the potential

$$U(q) = \frac{1}{4}q^4 + \frac{\mu_0}{3}q^3 + \frac{\mu_1}{2}q^2 + \mu_2 q \qquad (1)$$

where μ_0, μ_1 and μ_2 are real parameters. Show that the term $\mu_0 q^3/3$ can be eliminated with the help of the transformation

$$\widetilde{q}(q) = q + \delta. \qquad (2)$$

(ii) Let

$$U(q, \mu_1, \mu_2) = \frac{1}{4}q^4 + \frac{\mu_1}{2}q^2 + \mu_2 q.$$

Find the parameter region where the potential $U(q) = q^4/4$ changes qualitatively with respect to μ_1 and μ_2.
(iii) Show that the *van der Waals equation*

$$(V - b)\left(P + \frac{a}{V^2}\right) = RT$$

can be expressed as

$$q^3 + \mu_1 q + \mu_2 = 0$$

where a, b and R are positive constants. Here P, V and T are the pressure, volume and temperature, respectively.

Solution 2. (i) From (2) we obtain

$$q^3 = \tilde{q}^3 - 3\delta\tilde{q}^2 + \cdots, \qquad q^4 = \tilde{q}^4 - 4\delta\tilde{q}^3 + \cdots.$$

Therefore the condition to eliminate the cubic term in U is $\mu_0 = 3\delta$.
(ii) The condition

$$\frac{\partial U}{\partial q} = 0$$

gives the cubic equation $q^3 + \mu_1 q + \mu_2 = 0$. This cubic equation defines a surface in the (q, μ_1, μ_2) space. The solution of the cubic equation is given by

$$q_1 = y_1 + y_2$$
$$q_2 = -\frac{y_1 + y_2}{2} + \frac{y_1 - y_2}{2} i\sqrt{3}$$
$$q_3 = -\frac{y_1 + y_2}{2} - \frac{y_1 - y_2}{2} i\sqrt{3}$$

where

$$y_1 := \sqrt[3]{-\frac{\mu_2}{2} + \sqrt{D}}, \qquad y_2 := \sqrt[3]{-\frac{\mu_2}{2} - \sqrt{D}}.$$

Here

$$D := \left(\frac{\mu_2}{2}\right)^2 + \left(\frac{\mu_1}{3}\right)^3$$

is the *discriminant*. Since μ_1 and μ_2 are real we find (a) one root is real and the two others are complex conjugate if $D > 0$, (b) all roots are real and at least two are equal if $D = 0$, (c) all roots are real and unequal if $D < 0$. We now have to investigate whether the real solutions lead to maxima, minima or points of inflexion for the potential (3). If $D > 0$, the potential (3) has one minimum at $q = 0$. There is no qualitative change of the potential $U(q) = q^4/4$. If $D < 0$, the potential (3) has two minima symmetric to $q = 0$. The point $q = 0$ is a local maximum of the potential U. In this case we have a qualitative change of the potential (3). Consequently, $D = 0$ is a bifurcation line in the μ_1-μ_2 plane. This problem is the so-called *cusp catastrophe* in catastrophe theory.
(iii) The van der Waals equation can be written near the critical point as $(V - V_c)^3 = 0$ or

$$V^3 - 3V_c V^2 + 3V_c^2 V - V_c^3 = 0$$

where V_c is the critical pressure. This equation is compared with the van der Waals equation with $T = T_c$ and $P = P_c$, i.e.,

$$(V - b)\left(P_c + \frac{a}{V^2}\right) = RT_c.$$

From this equation it follows that

$$V^3 - \left(b + \frac{RT_c}{P_c}\right)V^2 + \frac{a}{P_c}V - \frac{ab}{P_c} = 0.$$

Comparing the coefficients we obtain

$$3V_c = b + \frac{RT_c}{P_c}, \qquad 3V_c^2 = \frac{a}{P_c}, \qquad V_c^3 = \frac{ab}{P_c}.$$

The solution of this system of equations is given by

$$RT_c = \frac{8a}{27b}, \qquad P_c = \frac{a}{27b^2}, \qquad V_c = 3b.$$

We now introduce the normalized quantities

$$\bar{P} := \frac{P}{P_c}, \qquad \bar{T} := \frac{T}{T_c}, \qquad \bar{V} := \frac{V}{V_c}.$$

We find

$$\left(\bar{P} + \frac{3}{\bar{V}^2}\right)\left(\bar{V} - \frac{1}{3}\right) = \frac{8}{3}\bar{T}.$$

Introducing the density $\bar{X} := \frac{1}{\bar{V}}$ leads to

$$(\bar{P} + 3\bar{X}^2)\left(\frac{1}{\bar{X}} - \frac{1}{3}\right) = \frac{8}{3}\bar{T}.$$

With the transformation $p := \bar{P} - 1$, $x := \bar{X} - 1$, $t := \bar{T} - 1$ we obtain

$$x^3 + \frac{1}{3}(8t + p)x + \frac{1}{3}(8t - 2p) = 0.$$

Therefore

$$x^3 + \mu_1 x + \mu_2 = 0$$

where

$$\mu_1 := \frac{1}{3}(8t + p), \qquad \mu_2 := \frac{1}{3}(8t - 2p).$$

Problem 3.   Let $\mathbb{N}$ be the natural numbers. A set is called *denumerable* if it is equipotent to the set of natural numbers $\mathbb{N}$. Show that the set $\mathbb{N} \times \mathbb{N}$ is denumerable. This means find a $1 - 1$ map between $\mathbb{N}$ and $\mathbb{N} \times \mathbb{N}$.

Solution 3.   We write the elements of $\mathbb{N} \times \mathbb{N}$ in the form of an array as follows

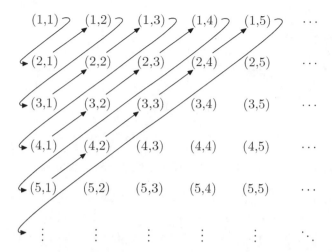

From the figure we see that we can arrange the elements of $\mathbb{N} \times \mathbb{N}$ into a (linear) sequence as indicated by the arrows, i.e.

$$(1,1),\ (2,1),\ (1,2),\ (3,1),\ (2,2),\ (1,3),\ (4,1),\ (3,2), \ldots.$$

Thus we find a $1-1$ map between $\mathbb{N}$ and $\mathbb{N} \times \mathbb{N}$. If $(m,n) \in \mathbb{N} \times \mathbb{N}$ we have

$$f(m,n) = \frac{1}{2}(m+n-1)(m+n-2)+n. \tag{1}$$

Equation (1) can be found as follows: The pair (m,n) lies in the $(m+n-1)$-th diagonal stripe in the above figure and is the n-th pair up from the left on that stripe. In the first stripe there is only one pair, in the second there are two pairs, and so on. Thus (m,n) occurs in position numbered

$$n + \sum_{k=1}^{m+n-2} k$$

in the counting procedure. Thus (1) follows. How do you find m and n if $f(m,n)$ is given?

Problem 4.   Consider the system of polynomial equations

$$\begin{aligned}
f_1(\mathbf{x}) &= x_1 - x_2 - x_3 = 0 \\
f_2(\mathbf{x}) &= x_1 + x_2 - x_3^2 = 0 \\
f_3(\mathbf{x}) &= x_1^2 + x_2^2 - 1 = 0.
\end{aligned}$$

Use a *Gaussian elimination-like method* to find the solutions of this system of polynomial equations.

Solution 4.   We choose x_1 in the first polynomial as the first term for eliminating terms in the two other polynomials.

$$g_1(\mathbf{x}) = f_1(\mathbf{x}) = x_1 - x_2 - x_3$$
$$g_2(\mathbf{x}) = f_2(\mathbf{x}) - f_1(\mathbf{x}) = 2x_2 - x_3^2 + x_3$$
$$g_3(\mathbf{x}) = f_3(\mathbf{x}) - (x_1 + x_2 + x_3)f_1(\mathbf{x}) = 2x_2^2 + 2x_2x_3 + x_3^2 - 1.$$

We choose the variable x_2 in g_2 as the most important variable. Then we multiply g_2 by another polynomial and subtract it from $2g_2$ in order to eliminate the terms containing x_2. We do the same for the third polynomial

$$h_1(\mathbf{x}) = 2g_1(\mathbf{x}) + g_2(\mathbf{x}) = 2x_1 - x_3^2 - x_3$$
$$h_2(\mathbf{x}) = g_2(\mathbf{x}) = 2x_2 - x_3^2 + x_3$$
$$h_3(\mathbf{x}) = 2g_3(\mathbf{x}) - (2x_2 + x_3^2 + x_3)g_2(\mathbf{x}) = x_3^4 + x_3^2 - 2.$$

The new set of equations is in upper triangular-form. The last polynomial is only in x_3, the second one is only in x_2 and x_3, and the first one is a polynomial in x_1 and x_3.

Problem 5.   (i) Find the maximum area of all triangles that can be inscribed in an ellipse

$$\frac{x^2}{a^2} + \frac{y^2}{b^2} = 1$$

with semiaxes a and b.
(ii) Describe the triangles that have maximum area.

Solution 5.   (i) We represent the ellipse using the parametric representation

$$x(t) = a\cos(t), \qquad y(t) = b\sin(t)$$

where $t \in [0, 2\pi)$. A triple of points on the ellipse is given by

$$(a\cos t_j, b\sin t_j), \quad j = 1, 2, 3.$$

Thus the area A of an inscribed triangle is given by

$$A = \frac{1}{2}\det\begin{pmatrix} 1 & a\cos t_1 & b\sin t_1 \\ 1 & a\cos t_2 & b\sin t_2 \\ 1 & a\cos t_3 & b\sin t_3 \end{pmatrix} = \frac{ab}{2}\det\begin{pmatrix} 1 & \cos t_1 & \sin t_1 \\ 1 & \cos t_2 & \sin t_2 \\ 1 & \cos t_3 & \sin t_3 \end{pmatrix}.$$

This is ab times the area of a triangle inscribed in the unit circle. Hence, the area is maximal if

$$t_2 = t_1 + \frac{2\pi}{3} \quad \text{and} \quad t_3 = t_2 + \frac{2\pi}{3}.$$

The area A is given by

$$A = \frac{3\sqrt{3}}{4}ab$$

where we used the fact that $\sin(2\pi/3) = \sqrt{3}/2$ and $\cos(2\pi/3) = -1/2$.
(ii) Thus the area is maximal when the corresponding triangle inscribed in the unit circle is regular.

Problem 6. (i) Consider the *Lorentz transformation*

$$x' = \frac{1}{\sqrt{1-\beta^2}}(x - vt), \qquad t' = \frac{1}{\sqrt{1-\beta^2}}\left(t - \frac{v}{c^2}x\right)$$

where $\beta = v/c$. Assume that

$$k'_x x' - \omega' t' = k_x x - \omega t$$

where ω is the frequency and $\mathbf{k} = (k_x, k_y, k_z)$ is the wave vector. Find the transformation between k'_x, ω' and k_x, ω.
(ii) Consider the special case $\omega = ck_x$ for the transformation.
(iii) Simplify the transformation for $v/c \ll 1$.

Solution 6. (i) Inserting the Lorentz transformation into $k'_x x' - \omega' t'$ yields

$$k'_x x' - \omega' t' = k'_x \frac{1}{\sqrt{1-\beta^2}}(x - vt) - \omega' \frac{1}{\sqrt{1-\beta^2}}\left(t - \frac{v}{c^2}x\right)$$

$$= \left(\frac{k'_x}{\sqrt{1-\beta^2}} + \frac{\omega' v}{\sqrt{1-\beta^2}c^2}\right)x - \left(\frac{vk'_x}{\sqrt{1-\beta^2}} + \frac{\omega'}{\sqrt{1-\beta^2}}\right)t$$

$$= k_x x - \omega t.$$

Comparing with respect to x and t provides the transformation

$$k_x = \frac{1}{\sqrt{1-\beta^2}}\left(k'_x + \frac{v}{c^2}\omega'\right)$$

$$\omega = \frac{1}{\sqrt{1-\beta^2}}(vk'_x + \omega').$$

The inverse transformation in matrix notation is

$$\begin{pmatrix} k'_x \\ \omega' \end{pmatrix} = \frac{1}{\sqrt{1-\beta^2}} \begin{pmatrix} 1 & -v/c^2 \\ -v & 1 \end{pmatrix} \begin{pmatrix} k_x \\ \omega \end{pmatrix}.$$

(ii) With the assumption $\omega = ck_x$ the transformation reduces to

$$\omega = \frac{\omega'}{\sqrt{1-\beta^2}}\left(1 + \frac{v}{c}\right).$$

(iii) Since $v \ll c$ and $\beta^2 = v^2/c^2$ we have $\sqrt{1 - \beta^2} \approx 1$. Therefore

$$\omega = \omega' \left(1 + \frac{v}{c}\right).$$

Problem 7. Let

$$S^2 := \{ (x_1, x_2, x_3) \in \mathbb{R}^3 \; : \; x_1^2 + x_2^2 + x_3^2 = 1 \}.$$

An element $\eta \in S^2$ can be written as

$$\eta = (\cos \phi \sin \theta, \sin \phi \sin \theta, \cos \theta)$$

where $\phi \in [0, 2\pi)$ and $\theta \in [0, \pi]$. The *stereographic projection* is a map

$$\Pi \; : \; S^2 \setminus \{ (0, 0, -1) \} \to \mathbb{R}^2$$

given by

$$x_1(\theta, \phi) = \frac{2 \sin(\theta) \cos(\phi)}{1 + \cos(\theta)}, \qquad x_2(\theta, \phi) = \frac{2 \sin(\theta) \sin(\phi)}{1 + \cos(\theta)}.$$

(i) Let $\theta = 0$ and ϕ be arbitrary. Find x_1, x_2. Give a geometric interpretation.
(ii) Find the inverse of the map, i.e., find

$$\Pi^{-1} \; : \; \mathbb{R}^2 \to S^2 \setminus \{ (0, 0, -1) \}.$$

Solution 7.   (i) Since $\sin(0) = 0$ we find $x_1 = x_2 = 0$, i.e. the point $(0, 0, 1)$ is mapped to the origin $(0, 0)$.
(ii) Using division we find

$$\phi(x_1, x_2) = \arctan \left(\frac{x_2}{x_1}\right).$$

Since

$$x_1^2 + x_2^2 = \frac{4 \sin^2 \theta}{(1 + \cos \theta)^2}$$

and

$$\tan \left(\frac{\theta}{2}\right) = \frac{\sin \theta}{1 + \cos \theta}$$

we obtain

$$\theta(x_1, x_2) = 2 \arctan \left(\frac{\sqrt{x_1^2 + x_2^2}}{2}\right).$$

Problem 8. Let

$$S^n := \{ (x_1, x_2, \ldots, x_{n+1}) \in \mathbb{R}^{n+1} : x_1^2 + x_2^2 + \cdots + x_{n+1}^2 = 1 \}.$$

(i) Show that S^3 can be considered as a subset of $\mathbb{C}^2$ ($\mathbb{C}^2 \cong \mathbb{R}^4$)

$$S^3 = \{ (z_1, z_2) \in \mathbb{C}^2 : |z_1|^2 + |z_2|^2 = 1 \}.$$

(ii) The *Hopf map* $\pi : S^3 \to S^2$ is defined by

$$\pi(z_1, z_2) := (\bar{z}_1 z_2 + \bar{z}_2 z_1, -i\bar{z}_1 z_2 + i\bar{z}_2 z_1, |z_1|^2 - |z_2|^2).$$

Find the parametrization of S^3, i.e. find $z_1(\theta, \phi)$, $z_2(\theta, \phi)$ and thus show that indeed π maps S^3 onto S^2.
(iii) Show that $\pi(z_1, z_2) = \pi(z_1', z_2')$ if and only if $z_j' = e^{i\alpha} z_j$ ($j = 1, 2$) and $\alpha \in \mathbb{R}$.

Solution 8. (i) Let $z_1 = x_1 + iy_1$ and $z_2 = x_2 + iy_2$, where $x_1, x_2, y_1, y_2 \in \mathbb{R}$. Then from $|z_1|^2 + |z_2|^2 = 1$ it follows that

$$x_1^2 + y_1^2 + x_2^2 + y_2^2 = 1.$$

(ii) Since $|z_1|^2 + |z_2|^2 = 1$ we have the parametrization

$$z_1(\theta, \phi) = \cos(\theta/2)e^{i\phi_1}, \qquad z_2(\theta, \phi) = \sin(\theta/2)e^{i\phi_2}$$

where $0 \leq \theta \leq \pi$ and $\phi_1, \phi_2 \in \mathbb{R}$. Thus

$$\pi(\cos(\theta/2)e^{i\phi_1}, \sin(\theta/2)e^{i\phi_2}) = (\sin\theta\cos(\phi_2 - \phi_1), \sin\theta\sin(\phi_2 - \phi_1), \cos\theta).$$

(iii) From $\pi(z_1, z_2) = \pi(z_1', z_2')$ we obtain the three equations

$$\bar{z}_1 z_2 + \bar{z}_2 z_1 = \bar{z}_1' z_2' + \bar{z}_2' z_1'$$
$$-\bar{z}_1 z_2 + \bar{z}_2 z_1 = -\bar{z}_1' z_2' + \bar{z}_2' z_1'$$
$$|z_1|^2 - |z_2|^2 = |z_1'|^2 - |z_2'|^2.$$

Inserting $|z_1|^2 + |z_2|^2 = |z_1'|^2 + |z_2'|^2 = 1$ into the last equation yields $|z_1| = |z_1'|$ and $|z_2| = |z_2'|$. Adding the first two equations provides $z_1\bar{z}_2 = z_1'\bar{z}_2'$. Thus we obtain the solution $z_1' = e^{i\alpha} z_1$ and $z_2' = e^{i\alpha} z_2$, where $\alpha \in \mathbb{R}$.

Problem 9. The n-dimensional complex projective space $\mathbb{C}P^n$ is the set of all complex lines on $\mathbb{C}^{n+1}$ passing through the origin. Let f be the map that takes nonzero vectors in $\mathbb{C}^2$ to vectors in $\mathbb{R}^3$ by

$$f(z_1, z_2) = \left(\frac{z_1\bar{z}_2 + \bar{z}_1 z_2}{z_1\bar{z}_1 + \bar{z}_2 z_2}, \frac{z_1\bar{z}_2 - \bar{z}_1 z_2}{i(z_1\bar{z}_1 + \bar{z}_2 z_2)}, \frac{z_1\bar{z}_1 - \bar{z}_2 z_2}{z_1\bar{z}_1 + \bar{z}_2 z_2} \right).$$

The map f defines a bijection between $\mathbf{CP}^1$ and the unit sphere in $\mathbb{R}^3$. Consider the normalized vectors in $\mathbb{C}^2$

$$\begin{pmatrix} 1 \\ 0 \end{pmatrix}, \quad \begin{pmatrix} 0 \\ 1 \end{pmatrix}, \quad \frac{1}{\sqrt{2}}\begin{pmatrix} 1 \\ 1 \end{pmatrix}, \quad \frac{1}{\sqrt{2}}\begin{pmatrix} 1 \\ -1 \end{pmatrix}, \quad \frac{1}{\sqrt{2}}\begin{pmatrix} i \\ -i \end{pmatrix}.$$

Apply f to these vectors in $\mathbb{C}^2$.

Solution 9. Since all vectors in $\mathbb{C}^2$ are normalized we have $z_1\bar{z}_1 + z_2\bar{z}_2 = 1$. We find

$$\begin{pmatrix} 0 \\ 0 \\ 1 \end{pmatrix}, \quad \begin{pmatrix} 0 \\ 0 \\ -1 \end{pmatrix}, \quad \begin{pmatrix} 1 \\ 0 \\ 0 \end{pmatrix}, \quad \begin{pmatrix} -1 \\ 0 \\ 0 \end{pmatrix}, \quad \begin{pmatrix} -1 \\ 0 \\ 0 \end{pmatrix}.$$

Problem 10. Consider the second order linear differential equation

$$\frac{d^2\phi}{dt^2} = 0. \tag{1}$$

(i) Let

$$u(t) = \frac{d\ln\phi}{dt}. \tag{2a}$$

Derive the differential equation for u.
(ii) Let

$$u(t) = \frac{d^2\ln\phi}{dt^2}. \tag{2b}$$

Derive the differential equation for u.

Hint. Use the following notation

$$A_1(t) := \int^t u(t_0)dt_0, \qquad A_2(t) := \int^t dt_1 \int^{t_1} u(t_0)dt_0$$

Solution 10. (i) Taking the derivative of A_2 and A_1 yields

$$\frac{dA_2}{dt} = A_1, \qquad \frac{dA_1}{dt} = A_0 \equiv u$$

and

$$\frac{de^{A_n}}{dt} = A_{n-1}e^{A_n}$$

where $n = 1, 2$. From (2a) we obtain

$$u(t) = \frac{1}{\phi}\frac{d\phi}{dt}.$$

Therefore $\phi(t) = \exp(A_1)$. Taking the derivative we find

$$\frac{d\phi}{dt} = e^{A_1} u$$

and

$$\frac{d^2\phi}{dt^2} = e^{A_1} u^2 + e^{A_1}\frac{du}{dt} \equiv e^{A_1}\left(\frac{du}{dt} + u^2\right).$$

Consequently (2a) leads to the nonlinear differential equation

$$\frac{du}{dt} + u^2 = 0.$$

This is a *Riccati equation*.

(ii) From (2b) we obtain $\phi(t) = \exp(A_2)$. The derivatives of ϕ give

$$\frac{d\phi}{dt} = e^{A_2} A_1$$

and

$$\frac{d^2\phi}{dt^2} = e^{A_2}(A_1^2 + u).$$

Therefore

$$A_1^2 + u = 0.$$

Taking the derivative of this equation leads to

$$2A_1 u + \frac{du}{dt} = 0$$

and

$$2u^2 + 2A_1\frac{du}{dt} + \frac{d^2u}{dt^2} = 0.$$

To eliminate A_1 we multiply this equation with u and insert $2A_1 u + du/dt = 0$. This leads to the nonlinear differential equation

$$u\frac{d^2u}{dt^2} - \left(\frac{du}{dt}\right)^2 + 2u^3 = 0.$$

The technique to find nonlinear ordinary differential equations from a linear differential equation using (1) can be extended to

$$A_n(t) := \int^t dt_{n-1}\int^{t_{n-1}} dt_{n-2}\cdots\int^{t_1} u(t_0)dt_0.$$

Problem 11.   Let

$$\frac{d^2w}{dz^2} + p_1(z)\frac{dw}{dz} + p_2(z)w = 0 \tag{1}$$

be a second-order linear differential equation in the complex domain. Consider the transformation

$$\tilde{z}(z) = z$$

$$v(\tilde{z}(z)) = w(z) \exp\left(\frac{1}{2}\int^z p_1(s)ds\right).$$

Find the differential equation for $v(\tilde{z})$.

Solution 11.   Let

$$R(z) := \frac{1}{2}\int^z p_1(s)ds.$$

Then

$$\frac{dv}{dz} = \frac{dv}{d\tilde{z}}\frac{d\tilde{z}}{dz} = \frac{dv}{d\tilde{z}} = \frac{dw}{dz}e^R + \frac{1}{2}p_1 w e^R \tag{1}$$

where we have used that $dR/dz = p_1/2$. Since $d\tilde{z}/dz = 1$ we obtain from (1)

$$\frac{d^2v}{d\tilde{z}^2} = \frac{d^2w}{dz^2}e^R + p_1\frac{dw}{dz}e^R + \frac{1}{2}\frac{dp_1}{dz}w e^R + \frac{1}{4}p_1^2 w e^R.$$

Consequently

$$\frac{d^2v}{d\tilde{z}^2} + I(\tilde{z})v = 0$$

where

$$I(\tilde{z}) := p_2(\tilde{z}) - \frac{1}{2}\frac{dp_1(\tilde{z})}{d\tilde{z}} - \frac{1}{4}p_1^2(\tilde{z}).$$

Problem 12.   The nonlinear ordinary differential equation

$$\frac{d^2u}{dt^2} + 3\frac{du}{dt}u + u^3 = 0 \tag{1}$$

occurs in the investigation of univalued functions defined by second-order differential equations and in the study of the *modified Emden equation*.
(i) Show that (1) can be linearized with the help of the transformations

$$U(T) = u^2(t(T)), \qquad dT = u(t(T))dt(T) \tag{2}$$

$$u(t) = \frac{1}{\phi(t)}\frac{d\phi(t)}{dt}. \tag{3}$$

(ii) Find the general solution to (1).

Solution 12.   (i) First we consider the transformation given by (2). Since

$$\frac{dU}{dT} = 2u\frac{du}{dt}\frac{dt}{dT} = 2\frac{du}{dt}$$

and

$$\frac{d^2U}{dT^2} = 2\frac{d^2u}{dt^2}\frac{dt}{dT} = \frac{2}{u}\frac{d^2u}{dt^2}$$

we obtain

$$\frac{u}{2}\frac{d^2U}{dT^2} + \frac{3u}{2}\frac{dU}{dT} + uU = 0.$$

With $u \neq 0$ it follows that

$$\frac{d^2U}{dT^2} + 3\frac{dU}{dT} + 2U = 0.$$

Now we consider the transformation given by (3). Since

$$\frac{du}{dt} = \frac{1}{\phi^2}\left(\frac{d^2\phi}{dt^2}\phi - \left(\frac{d\phi}{dt}\right)^2\right)$$

and

$$\frac{d^2u}{dt^2} = \frac{1}{\phi^3}\left(\frac{d^3\phi}{dt^3}\phi^2 - 3\phi\frac{d^2\phi}{dt^2}\frac{d\phi}{dt} + 2\left(\frac{d\phi}{dt}\right)^3\right)$$

we obtain (with $\phi \neq 0$) the linear differential equation

$$\frac{d^3\phi}{dt^3} = 0. \tag{4}$$

(ii) The general solution to (1) can now easily be found with the help of (4) and transformation (2). The general solution to (4) is given by

$$\phi(t) = C_1 t^2 + C_2 t + C_3$$

where C_1, C_2 and C_3 are the constants of integration. Inserting $\phi(t)$ into (3) yields the general solution

$$u(t) = \frac{2t + (C_2/C_1)}{t^2 + (C_2/C_1)t + (C_3/C_1)}$$

where $C_1 \neq 0$. If $C_1 = 0$ and $C_2 \neq 0$, then

$$u(t) = \frac{1}{t + C_3/C_2}.$$

If $C_1 = C_2 = 0$ and $C_3 \neq 0$, then $u(t) = 0$.

Problem 13. Consider the ordinary differential equation

$$\frac{d^2u}{dt^2} + \lambda(\phi(t))^{2m-2}u = 0 \tag{1}$$

where $m = 1, 2, \ldots$, and λ is a real parameter The smooth function ϕ satisfies

$$\frac{d^2\phi}{dt^2} + (\phi(t))^{2m-1} = 0 \tag{2}$$

with the initial condition $\phi(0) = 1$, $d\phi(0)/dt = 0$. Perform the transformation

$$z(t) = (\phi(t))^{2m}, \qquad \bar{u}(z(t)) = u(t).$$

Solution 13.   First we notice that the integration of the differential equation (2) yields

$$\frac{1}{2}\left(\frac{d\phi}{dt}\right)^2 + \frac{1}{2m}\phi^{2m} = \frac{1}{2m}$$

where we have taken into account the initial conditions. From the transformation it follows that

$$\frac{d\bar{u}}{dt} = \frac{d\bar{u}}{dz}\frac{dz}{dt} = \frac{du}{dt}$$

and

$$\frac{d^2\bar{u}}{dt^2} = \frac{d^2\bar{u}}{dz^2}\left(\frac{dz}{dt}\right)^2 + \frac{d\bar{u}}{dz}\frac{d^2z}{dt^2} = \frac{d^2u}{dt^2}.$$

From $z(t) = (\phi(t))^m$ we find that

$$\frac{dz}{dt} = 2m(\phi(t))^{2m-1}\frac{d\phi}{dt}.$$

Consequently,

$$\left(\frac{dz}{dt}\right)^2 = 4m^2(\phi(t))^{4m-2}\left(\frac{d\phi}{dt}\right)^2 = 4m(\phi(t))^{2m-2}z(1-z).$$

From this equation we also obtain

$$\frac{d^2z}{dt^2} = 2(\phi(t))^{2m-2}((1-3m)z + 2m - 1).$$

It follows that

$$z(1-z)\frac{d^2\bar{u}}{dz^2} + (\gamma - (\alpha + \beta + 1)z)\frac{d\bar{u}}{dz} - \alpha\beta\bar{u} = 0$$

where

$$\alpha + \beta = \frac{m-1}{2m}, \qquad \alpha\beta = \frac{-\lambda}{4m}, \qquad \gamma = \frac{2m-1}{2m}.$$

This second order linear ordinary differential equation has three singular points at $z = 0$, $z = 1$ and $z = \infty$. It is called the *hypergeometric equation*.

Problem 14. Show that the one-dimensional *wave equation*

$$\frac{1}{c^2}\frac{\partial^2 u}{\partial t^2} = \frac{\partial^2 u}{\partial x^2} \tag{1}$$

is invariant under the transformation

$$\begin{pmatrix} x'(x,t) \\ ct'(x,t) \end{pmatrix} = \begin{pmatrix} \cosh\epsilon & \sinh\epsilon \\ \sinh\epsilon & \cosh\epsilon \end{pmatrix} \begin{pmatrix} x \\ ct \end{pmatrix} \tag{2a}$$

$$u'(x'(x,t),t'(x,t)) = u(x,t) \tag{2b}$$

where ϵ is a real parameter. Equation (2a) is called the *Lorentz transformation*. We have to show that

$$\frac{1}{c^2}\frac{\partial^2 u'}{\partial t'^2} = \frac{\partial^2 u'}{\partial x'^2}. \tag{3}$$

Solution 14. Applying the *chain rule* gives

$$\frac{\partial u'}{\partial t} = \frac{\partial u'}{\partial x'}\frac{\partial x'}{\partial t} + \frac{\partial u'}{\partial t'}\frac{\partial t'}{\partial t} = \frac{\partial u}{\partial t}$$

and

$$\frac{\partial u'}{\partial x} = \frac{\partial u'}{\partial x'}\frac{\partial x'}{\partial x} + \frac{\partial u'}{\partial t'}\frac{\partial t'}{\partial x} = \frac{\partial u}{\partial x}.$$

It follows that

$$\frac{\partial^2 u'}{\partial t^2} = \left(\frac{\partial^2 u'}{\partial x'^2}\frac{\partial x'}{\partial t} + \frac{\partial^2 u'}{\partial x'\partial t'}\frac{\partial t'}{\partial t}\right)\frac{\partial x'}{\partial t} + \left(\frac{\partial^2 u'}{\partial x'\partial t'}\frac{\partial x'}{\partial t} + \frac{\partial^2 u'}{\partial t'^2}\frac{\partial t'}{\partial t}\right)\frac{\partial t'}{\partial t} = \frac{\partial^2 u}{\partial t^2}$$

and

$$\frac{\partial^2 u'}{\partial x^2} = \left(\frac{\partial^2 u'}{\partial x'^2}\frac{\partial x'}{\partial x} + \frac{\partial^2 u'}{\partial x'\partial t'}\frac{\partial t'}{\partial x}\right)\frac{\partial x'}{\partial x} + \left(\frac{\partial^2 u'}{\partial x'\partial t'}\frac{\partial x'}{\partial x} + \frac{\partial^2 u'}{\partial t'^2}\frac{\partial t'}{\partial x}\right)\frac{\partial t'}{\partial x} = \frac{\partial^2 u}{\partial x^2}.$$

Since

$$\frac{\partial x'}{\partial t} = c\sinh\epsilon, \qquad \frac{\partial x'}{\partial x} = \cosh\epsilon$$

and

$$\frac{\partial t'}{\partial t} = \cosh\epsilon, \qquad \frac{\partial t'}{\partial x} = \frac{1}{c}\sinh\epsilon$$

we obtain

$$c^2\frac{\partial^2 u'}{\partial x'^2}\sinh^2\epsilon + 2c\frac{\partial^2 u'}{\partial x'\partial t'}\sinh\epsilon\cosh\epsilon + \frac{\partial^2 u'}{\partial t'^2}\cosh^2\epsilon = \frac{\partial^2 u}{\partial t^2}$$

and

$$\frac{\partial^2 u'}{\partial x'^2}\cosh^2\epsilon + \frac{2}{c}\frac{\partial^2 u'}{\partial x'\partial t'}\sinh\epsilon\cosh\epsilon + \frac{1}{c^2}\frac{\partial^2 u'}{\partial t'^2}\sinh^2\epsilon = \frac{\partial^2 u}{\partial x^2}.$$

Inserting these expressions into the wave equation (1) we find (3), where we have used the identity $\cosh^2 \epsilon - \sinh^2 \epsilon \equiv 1$ for all $\epsilon \in \mathbb{R}$. Thus the partial differential equation (1) is *invariant* under the transformation (2).

Problem 15.   Consider the nonlinear partial differential equation

$$\left(\frac{\partial u}{\partial t}\right)^2 \frac{\partial^2 u}{\partial x^2} + \left(\frac{\partial u}{\partial x}\right)^2 \frac{\partial^2 u}{\partial t^2} - 2\frac{\partial u}{\partial x}\frac{\partial u}{\partial t}\frac{\partial^2 u}{\partial x \partial t} = 0. \tag{1}$$

(i) Show that the partial differential equation (1) can be linearized with the following transformation

$$\epsilon(x,t) = \frac{\partial u}{\partial x} \tag{2a}$$

$$\eta(x,t) = \frac{\partial u}{\partial t} \tag{2b}$$

$$x(\epsilon,\eta) = \frac{\partial W}{\partial \epsilon} \tag{2c}$$

$$t(\epsilon,\eta) = \frac{\partial W}{\partial \eta} \tag{2d}$$

$$u(x,t) + W(\epsilon(x,t), \eta(x,t)) = x\epsilon(x,t) + t\eta(x,t). \tag{2e}$$

The transformation (2) is called the *Legendre transformation*.
(ii) Give the solution to the linearized partial differential equation.

Solution 15.   (i) Since

$$\epsilon = \frac{\partial u}{\partial x}(x(\epsilon,\eta), t(\epsilon,\eta))$$

we obtain by taking the derivative with respect to ϵ

$$1 = \frac{\partial^2 u}{\partial x^2}\frac{\partial x}{\partial \epsilon} + \frac{\partial^2 u}{\partial x \partial t}\frac{\partial t}{\partial \epsilon} = \frac{\partial^2 u}{\partial x^2}\frac{\partial^2 W}{\partial \epsilon^2} + \frac{\partial^2 u}{\partial x \partial t}\frac{\partial^2 W}{\partial \epsilon \partial \eta}$$

where we have used (2c) and (2d). Analogously, we find

$$0 = \frac{\partial^2 u}{\partial x \partial t}\frac{\partial^2 W}{\partial \epsilon^2} + \frac{\partial^2 u}{\partial t^2}\frac{\partial^2 W}{\partial \epsilon \partial \eta}, \qquad 0 = \frac{\partial^2 u}{\partial x^2}\frac{\partial^2 W}{\partial \epsilon \partial \eta} + \frac{\partial^2 u}{\partial x \partial t}\frac{\partial^2 W}{\partial \eta^2}$$

and

$$1 = \frac{\partial^2 u}{\partial x \partial t}\frac{\partial^2 W}{\partial \epsilon \partial \eta} + \frac{\partial^2 u}{\partial t^2}\frac{\partial^2 W}{\partial \eta^2}.$$

If we set

$$\rho := \frac{\partial^2 u}{\partial x^2}\frac{\partial^2 u}{\partial t^2} - \left(\frac{\partial^2 u}{\partial x \partial t}\right)^2$$

we obtain

$$\frac{\partial^2 W}{\partial \epsilon^2} \frac{\partial^2 W}{\partial \eta^2} - \left(\frac{\partial^2 W}{\partial \epsilon \partial \eta} \right)^2 = \frac{1}{\rho}.$$

Inserting these expressions into (1) we obtain the linear partial differential equation

$$\epsilon^2 \frac{\partial^2 W}{\partial \epsilon^2} + 2\epsilon\eta \frac{\partial^2 W}{\partial \epsilon \partial \eta} + \eta^2 \frac{\partial^2 W}{\partial \eta^2} = 0.$$

(ii) Since this partial differential equation can be written in the form

$$\left(\left(\epsilon \frac{\partial}{\partial \epsilon} + \eta \frac{\partial}{\partial \eta} \right)^2 - \left(\epsilon \frac{\partial}{\partial \epsilon} + \eta \frac{\partial}{\partial \eta} \right) \right) W = 0$$

we find that the general solution is given by

$$W(\epsilon, \eta) = G\left(\frac{\eta}{\epsilon} \right) + \eta H \left(\frac{\epsilon}{\eta} \right)$$

where G and H are two arbitrary smooth functions.

Problem 16. Let

$$\bar{t}(x,t) = t, \qquad \bar{x}(x,t) = \int^x \frac{1}{u(s,t)} ds$$

$$\bar{u}(\bar{x}(x,t), \bar{t}(x,t)) = u(x,t).$$

Let

$$\frac{\partial u}{\partial t} = u^2 \frac{\partial^2 u}{\partial x^2}.$$

Find the partial differential equation for $\bar{u}(\bar{x}, \bar{t})$.

Solution 16. We obtain

$$\frac{\partial \bar{t}}{\partial t} = 1, \qquad \frac{\partial \bar{t}}{\partial x} = 0, \qquad \frac{\partial \bar{x}}{\partial x} = \frac{1}{u(x,t)}$$

and

$$\frac{\partial \bar{x}}{\partial t} = -\int^x \frac{u_t(s,t)}{u^2(s,t)} ds = -\int^x \frac{\partial^2 u(s,t)}{\partial s^2} ds = -\frac{\partial u}{\partial x}$$

where $u_t \equiv \partial u/\partial t$. Applying the chain rule we have

$$\frac{\partial \bar{u}}{\partial x} = \frac{\partial \bar{u}}{\partial \bar{x}} \frac{\partial \bar{x}}{\partial x} + \frac{\partial \bar{u}}{\partial \bar{t}} \frac{\partial \bar{t}}{\partial x} = \frac{\partial u}{\partial x}.$$

Therefore

$$\frac{\partial \bar{u}}{\partial \bar{x}} = u \frac{\partial u}{\partial x}.$$

Analogously,

$$\frac{\partial \bar{u}}{\partial t} = \frac{\partial \bar{u}}{\partial \bar{x}}\frac{\partial \bar{x}}{\partial t} + \frac{\partial \bar{u}}{\partial \bar{t}}\frac{\partial \bar{t}}{\partial t} = \frac{\partial u}{\partial t}$$

and therefore

$$-\frac{\partial \bar{u}}{\partial \bar{x}}\frac{\partial u}{\partial x} + \frac{\partial \bar{u}}{\partial \bar{t}} = \frac{\partial u}{\partial t}.$$

Furthermore we obtain

$$\frac{\partial}{\partial x}\left(\frac{\partial \bar{u}}{\partial \bar{x}}\right) = \left(\frac{\partial u}{\partial x}\right)^2 + u\frac{\partial^2 u}{\partial x^2}.$$

It follows that

$$\left(\frac{\partial^2 \bar{u}}{\partial \bar{x}^2}\frac{\partial \bar{x}}{\partial x} + \frac{\partial^2 \bar{u}}{\partial \bar{x}\partial \bar{t}}\frac{\partial \bar{t}}{\partial x}\right) = \left(\frac{\partial u}{\partial x}\right)^2 + u\frac{\partial^2 u}{\partial x^2}.$$

Since $\partial \bar{t}/\partial x = 0$ and $\partial \bar{x}/\partial x = u^{-1}$ we obtain

$$\frac{\partial^2 \bar{u}}{\partial \bar{x}^2} = u\left(\frac{\partial u}{\partial x}\right)^2 + u^2\frac{\partial^2 u}{\partial x^2}.$$

Finally we obtain

$$\frac{\partial \bar{u}}{\partial \bar{t}} = \frac{\partial^2 \bar{u}}{\partial \bar{x}^2}.$$

This partial differential equation is the *linear diffusion equation*.

Problem 17.   The nonlinear partial differential equation

$$\frac{\partial^2 u}{\partial t^2} - \frac{\partial^2 u}{\partial x^2} + f_1(x+t)f_2(x-t)\sin u = 0$$

is an extended one-dimensional *sine-Gordon equation*, where f_1 and f_2 are two smooth functions. Let

$$\xi(x,t) := \int^{x+t} f_1(s)ds \tag{1a}$$

$$\eta(x,t) := \int^{x-t} f_2(s)ds \tag{1b}$$

$$\bar{u}(\xi(x,t),\eta(x,t)) := u(x,t). \tag{1c}$$

Find the partial differential equation for $\bar{u}(\xi,\eta)$.

Solution 17.   From (1a) and (1b) we obtain

$$\frac{\partial \xi}{\partial x} = f_1(x+t), \qquad \frac{\partial \eta}{\partial x} = f_2(x-t) \tag{2a}$$

$$\frac{\partial \xi}{\partial t} = f_1(x+t), \qquad \frac{\partial \eta}{\partial t} = -f_2(x-t). \tag{2b}$$

Using the chain rule we have from (1c)

$$\frac{\partial \bar{u}}{\partial x} = \frac{\partial \bar{u}}{\partial \xi}\frac{\partial \xi}{\partial x} + \frac{\partial \bar{u}}{\partial \eta}\frac{\partial \eta}{\partial x} = \frac{\partial u}{\partial x}.$$

Thus

$$\frac{\partial \bar{u}}{\partial \xi} f_1(x+t) + \frac{\partial \bar{u}}{\partial \eta} f_2(x-t) = \frac{\partial u}{\partial x}. \tag{3}$$

Analogously,

$$\frac{\partial \bar{u}}{\partial t} = \frac{\partial \bar{u}}{\partial \xi}\frac{\partial \xi}{\partial t} + \frac{\partial \bar{u}}{\partial \eta}\frac{\partial \eta}{\partial t} = \frac{\partial u}{\partial t}$$

and

$$\frac{\partial \bar{u}}{\partial \xi} f_1(x+t) - \frac{\partial \bar{u}}{\partial \eta} f_2(x-t) = \frac{\partial u}{\partial t}. \tag{4}$$

It follows that

$$\frac{\partial}{\partial x}\left(\frac{\partial \bar{u}}{\partial \xi} f_1(x+t) + \frac{\partial \bar{u}}{\partial \eta} f_2(x-t)\right) = \frac{\partial^2 u}{\partial x^2}.$$

Therefore

$$\frac{\partial^2 \bar{u}}{\partial \xi^2} f_1^2 + 2\frac{\partial^2 \bar{u}}{\partial \xi \partial \eta} f_1 f_2 + \frac{\partial^2 \bar{u}}{\partial \eta^2} f_2^2 + \frac{\partial \bar{u}}{\partial \xi} f_1' + \frac{\partial \bar{u}}{\partial \eta} f_2' = \frac{\partial^2 u}{\partial x^2}$$

where $f_1' \equiv df_1(s)/ds$. Analogously, from (4) we obtain

$$\frac{\partial}{\partial t}\left(\frac{\partial \bar{u}}{\partial \xi} f_1(x+t) - \frac{\partial \bar{u}}{\partial \eta} f_2(x-t)\right) = \frac{\partial^2 u}{\partial t^2}.$$

Therefore

$$\frac{\partial^2 \bar{u}}{\partial \xi^2} f_1^2 - 2\frac{\partial^2 \bar{u}}{\partial \xi \partial \eta} f_1 f_2 + \frac{\partial^2 \bar{u}}{\partial \eta^2} f_2^2 + \frac{\partial \bar{u}}{\partial \xi} f_1' + \frac{\partial \bar{u}}{\partial \eta} f_2' = \frac{\partial^2 u}{\partial t^2}.$$

Consequently, we obtain

$$4\frac{\partial^2 \bar{u}}{\partial \eta \partial \xi} - \sin \bar{u} = 0.$$

Problem 18. Consider the transformation

$$\bar{t}(x,t) = t$$
$$\bar{x}(x,t) = u(x,t)$$
$$\bar{u}(\bar{x}(x,t), \bar{t}(x,t)) = x$$

and the nonlinear partial differential equation

$$\frac{\partial u}{\partial t} = \frac{\partial^2 u}{\partial x^2} + u \frac{\partial u}{\partial x}.$$

Find the partial differential equation for $\bar{u}(\bar{x}, \bar{t})$.

Solution 18.   Taking the derivatives with respect to t and x provides

$$\frac{\partial \bar{t}}{\partial t} = 1, \quad \frac{\partial \bar{t}}{\partial x} = 0, \quad \frac{\partial \bar{x}}{\partial t} = \frac{\partial u}{\partial t}, \quad \frac{\partial \bar{x}}{\partial x} = \frac{\partial u}{\partial x}.$$

and

$$\frac{\partial \bar{u}}{\partial x} = \frac{\partial \bar{u}}{\partial \bar{x}} \frac{\partial \bar{x}}{\partial x} + \frac{\partial \bar{u}}{\partial \bar{t}} \frac{\partial \bar{t}}{\partial x} = 1.$$

Therefore $(\partial \bar{u}/\partial \bar{x})(\partial u/\partial x) = 1$. Analogously,

$$\frac{\partial \bar{u}}{\partial t} = \frac{\partial \bar{u}}{\partial \bar{x}} \frac{\partial \bar{x}}{\partial t} + \frac{\partial \bar{u}}{\partial \bar{t}} \frac{\partial \bar{t}}{\partial t} = 0$$

and

$$\frac{\partial \bar{u}}{\partial \bar{x}} \frac{\partial u}{\partial t} + \frac{\partial \bar{u}}{\partial \bar{t}} = 0.$$

It follows that

$$\frac{\partial}{\partial x} \left(\frac{\partial \bar{u}}{\partial \bar{x}} \frac{\partial u}{\partial x} \right) = 0$$

and

$$\left(\frac{\partial^2 \bar{u}}{\partial \bar{x}^2} \frac{\partial \bar{x}}{\partial x} + \frac{\partial^2 \bar{u}}{\partial \bar{x} \partial \bar{t}} \frac{\partial \bar{t}}{\partial x} \right) \frac{\partial u}{\partial x} + \frac{\partial \bar{u}}{\partial \bar{x}} \frac{\partial^2 u}{\partial x^2} = 0.$$

Consequently

$$\frac{\partial^2 \bar{u}}{\partial \bar{x}^2} \left(\frac{\partial u}{\partial x} \right)^2 + \frac{\partial \bar{u}}{\partial \bar{x}} \frac{\partial^2 u}{\partial x^2} = 0.$$

Thus we have

$$u = \bar{x}, \quad \frac{\partial u}{\partial t} = -\frac{\bar{u}_{\bar{t}}}{\bar{u}_{\bar{x}}}, \quad \frac{\partial u}{\partial x} = \frac{1}{\bar{u}_{\bar{x}}}, \quad \frac{\partial^2 u}{\partial x^2} = -\frac{\bar{u}_{\bar{x}\bar{x}}}{\bar{u}_{\bar{x}}^3}$$

where $\bar{u}_{\bar{t}} := \partial \bar{u}/\partial \bar{t}$ etc. Inserting these results into the nonlinear partial differential equation provides the nonlinear partial differential equation

$$\left(\frac{\partial \bar{u}}{\partial \bar{x}} \right)^2 \frac{\partial \bar{u}}{\partial \bar{t}} = \frac{\partial^2 \bar{u}}{\partial \bar{x}^2} - \left(\frac{\partial \bar{u}}{\partial \bar{x}} \right)^2 \bar{x}.$$

Chapter 10

L'Hospital's Rule

Suppose that the functions g and h are continuously differentiable and $dh/dx \neq 0$ on an open interval I that contains a (except possibly at a). Suppose that

$$\lim_{x \to a} g(x) = 0 \quad \text{and} \quad \lim_{x \to a} h(x) = 0$$

or that

$$\lim_{x \to a} g(x) = \pm\infty \qquad \lim_{x \to a} h(x) = \pm\infty.$$

Thus with g/h we have an indeterminate form of type $0/0$ or ∞/∞. Then

$$\lim_{x \to a} \frac{g(x)}{h(x)} = \lim_{x \to a} \frac{dg(x)/dx}{dh(x)/dx}$$

if the limit on the right hand side exists (or is ∞ or $-\infty$). This is called *L'Hospital's rule.*

Problem 1. Consider

$$f(x) = \frac{\sin(x)}{x}$$

Determine $f(0)$ using L'Hospital's rule.

Solution 1. We have

$$\frac{d}{dx}\sin(x) = \cos(x), \qquad \frac{d}{dx}x = 1$$

and $\cos(0) = 1$. Thus $f(0) = 1$.

Problem 2.   (i) Let $f : \mathbb{R}^+ \to \mathbb{R}$

$$f(x) = x \ln(x) \quad \text{for} \quad x > 0.$$

Determine $f(0)$ using L'Hospital's rule.

Solution 2.   (i) Since

$$\lim_{x \to +0} x \ln x \equiv \lim_{x \to +0} \frac{\ln x}{1/x}$$

we have $g(x) = \ln x$ and $h(x) = 1/x$. It follows that

$$\frac{dg}{dx} = \frac{1}{x}, \qquad \frac{dh}{dx} = -\frac{1}{x^2}.$$

Therefore

$$\lim_{x \to +0} x \ln x = \lim_{x \to +0} \frac{\frac{1}{x}}{\frac{-1}{x^2}} = \lim_{x \to +0} \frac{-x^2}{x} = 0.$$

Problem 3.   Let

$$f(\theta) = \frac{\cos(\frac{\pi}{2} \cos \theta)}{\sin \theta} \quad \text{for} \quad \theta \neq n\pi \qquad (1)$$

where $n \in \mathbb{Z}$. Determine $f(n\pi)$ using L'Hospital's rule.

Solution 3.   Let

$$g(\theta) = \cos\left(\frac{\pi}{2} \cos \theta\right), \qquad h(\theta) = \sin \theta.$$

Then

$$\frac{dg}{d\theta} = \sin\left(\frac{\pi}{2} \cos \theta\right) \frac{\pi}{2} \sin \theta, \qquad \frac{dh}{d\theta} = \cos \theta.$$

Since

$$\left.\frac{dg}{d\theta}\right|_{\theta=n\pi} = 0, \qquad \left.\frac{dh}{d\theta}\right|_{\theta=n\pi} \neq 0$$

we obtain

$$\lim_{\theta \to n\pi} f(\theta) = 0.$$

Problem 4.   (i) Calculate

$$\lim_{\beta \to \infty} \tanh(\beta x)$$

where $x \in \mathbb{R}$.

(ii) Calculate

$$\lim_{\beta \to \infty} \frac{1}{\beta} \ln(4 \cosh(\beta x) \cosh(\beta y))$$

where $x, y \in \mathbb{R}$.

(iii) Find

$$\lim_{\beta \to \infty} \frac{\sinh(\beta x)}{\cosh(\beta x) + \cosh(\beta y)}$$

where $\beta > 0$.

Solution 4. (i) Let $x = 0$, then $\tanh(0) = 0$. If $x > 0$, then

$$\lim_{\beta \to \infty} \tanh(\beta x) = 1.$$

If $x < 0$, then

$$\lim_{\beta \to \infty} \tanh(\beta x) = -1.$$

Thus

$$\lim_{\beta \to \infty} \tanh(\beta x) = \begin{cases} 1 & x > 0 \\ 0 & x = 0 \\ -1 & x < 0. \end{cases}$$

One writes

$$\lim_{\beta \to \infty} \tanh(\beta x) = \operatorname{sgn}(x).$$

(ii) Applying L'Hospital's rule we have

$$\lim_{\beta \to \infty} \frac{1}{\beta} \ln[4 \cosh(\beta x) \cosh(\beta y)] = \lim_{\beta \to \infty} \frac{x \sinh(\beta x) \cosh(\beta y) + y \cosh(\beta x) \sinh(\beta y)}{\cosh(\beta x) \cosh(\beta y)}$$

$$= \lim_{\beta \to \infty} [x \tanh(\beta x) + y \tanh(\beta y)]$$

$$= x\operatorname{sgn}(x) + y\operatorname{sgn}(y)$$

$$= |x| + |y|.$$

(iii) Applying the addition theorem we find

$$\lim_{\beta \to \infty} \frac{\sinh(\beta x)}{\cosh(\beta x) + \cosh(\beta y)} = \lim_{\beta \to \infty} \frac{\sinh\left[\beta\left(\left(\frac{x}{2} + \frac{y}{2}\right) + \left(\frac{x}{2} - \frac{y}{2}\right)\right)\right]}{\cosh(\beta x) + \cosh(\beta y)}$$

$$= \frac{1}{2} \lim_{\beta \to \infty} \tanh\left(\beta \frac{x+y}{2}\right) + \frac{1}{2} \lim_{\beta \to \infty} \tanh\left(\beta \frac{x-y}{2}\right)$$

$$= \frac{1}{2}\operatorname{sgn}(x+y) + \frac{1}{2}\operatorname{sgn}(x-y).$$

where we have used the identity

$$\frac{\sinh\left[\beta\left(\left(\frac{x+y}{2}\right)+\left(\frac{x-y}{2}\right)\right)\right]}{\cosh(\beta x)+\cosh(\beta y)}\equiv$$

$$\frac{\sinh\left(\beta\frac{x+y}{2}\right)\cosh\left(\beta\frac{x-y}{2}\right)+\cosh\left(\beta\frac{x+y}{2}\right)\sinh\left(\beta\frac{x-y}{2}\right)}{2\cosh\left(\beta\frac{x+y}{2}\right)\cosh\left(\beta\frac{x-y}{2}\right)}.$$

Problem 5.   Calculate

$$\lim_{x\to y}\frac{\sin(x)-\sin(y)}{x-y}.$$

Solution 5.   We set $x=y+\epsilon$. Then

$$\lim_{\epsilon\to 0}\frac{\sin(y+\epsilon)-\sin(y)}{\epsilon}.$$

Since $\sin(y+\epsilon)\equiv\sin(y)\cos(\epsilon)+\cos(y)\sin(\epsilon)$ we have

$$\lim_{\epsilon\to 0}\frac{\sin(y)\cos(\epsilon)+\cos(y)\sin(\epsilon)-\sin(y)}{\epsilon}.$$

Applying L'Hospital's rule yields

$$\lim_{\epsilon\to 0}\frac{-\sin(y)\sin(\epsilon)+\cos(y)\cos(\epsilon)}{1}=\cos(y).$$

Chapter 11

Lagrange Multiplier Method

Let M be a smooth manifold and f be a real valued function of class $C^{(1)}$ on some open set containing M. We consider the problem of finding the extrema of the function $f|M$. This is called a problem of *constrained extrema*.

The *Lagrange multiplier rule* is as follows. Assume that f has a constrained relative extremum at $\mathbf{x}^* = (x_1^*, x_2^*, \dots, x_n^*)$. Then there exist real numbers $\lambda_1, \lambda_2, \dots, \lambda_m$ such that $\mathbf{x}^*$ is a critical point of the function

$$L(\mathbf{x}) := f(\mathbf{x}) + \lambda_1 g_1(\mathbf{x}) + \cdots + \lambda_m g_m(\mathbf{x}).$$

The numbers $\lambda_1, \lambda_2, \cdots, \lambda_m$ are called *Lagrange multipliers*. The critical points are given by the solution of the system

$$\frac{\partial L}{\partial x_k} = 0, \qquad k = 1, 2, \dots, n.$$

L is called the *Lagrangian*.

Problem 1. Let

$$f(\mathbf{x}) = x_1 - x_2 + 2x_3.$$

Find the maximum and minimum values of f on the *ellipsoid*

$$M := \{\,(x_1, x_2, x_3) : x_1^2 + x_2^2 + 2x_3^2 = 2\,\}.$$

Solution 1. Let

$$g(\mathbf{x}) = 2 - (x_1^2 + x_2^2 + 2x_3^2)$$

and

$$L(\mathbf{x}) = f(\mathbf{x}) + \lambda g(\mathbf{x}).$$

The Lagrange multiplier λ is yet to be determined. From the multiplier rule we obtain three equations

$$\left.\frac{\partial L}{\partial x_1}\right|_{\mathbf{x}=\mathbf{x}^*} = 1 - 2\lambda x_1^* = 0$$

$$\left.\frac{\partial L}{\partial x_2}\right|_{\mathbf{x}=\mathbf{x}^*} = -1 - 2\lambda x_2^* = 0$$

$$\left.\frac{\partial L}{\partial x_3}\right|_{\mathbf{x}=\mathbf{x}^*} = 2 - 4\lambda x_3^* = 0.$$

From these and the fourth equation (constraint) $g(\mathbf{x}^*) = 0$, we obtain

$$x_1^* = \frac{1}{2\lambda}, \qquad x_2^* = -\frac{1}{2\lambda}, \qquad x_3^* = \frac{1}{2\lambda}, \qquad \frac{1}{\lambda} = \pm\sqrt{2}.$$

Therefore

$$\mathbf{x}^* = \pm(\sqrt{2}/2)(\mathbf{e}_1 - \mathbf{e}_2 + \mathbf{e}_3)$$

depending on which of the two possible values for λ is used. Here $\mathbf{e}_1$, $\mathbf{e}_2$ and $\mathbf{e}_3$ denote the standard basis in $\mathbb{R}^3$. Since f is continuous and M is a compact set, f has a maximum and a minimum value on M. One of the two critical points obtained by the multiplier rule must give the maximum and the other the minimum. Since

$$f[(\sqrt{2}/2)(\mathbf{e}_1 - \mathbf{e}_2 + \mathbf{e}_3)] = 2\sqrt{2}$$

and

$$f[-(\sqrt{2}/2)(\mathbf{e}_1 - \mathbf{e}_2 + \mathbf{e}_3)] = -2\sqrt{2}$$

these numbers are the maximum and minimum values, respectively.

Problem 2. Calculate the greatest volume of a rectangular box that can be securely tied up with a piece of string 360 cm in length. The string must pass twice around the width of the parcel and once around the length as shown in the figure. Disregard the amount used up by the knot.

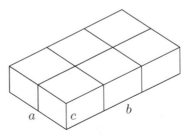

Solution 2. The problem can be solved using the Lagrange multiplier method. Let a, b, c be the width, length and height respectively of the box. The length of the string is

$$S = 2b + 4a + 6c = 360 \text{ cm}. \tag{1}$$

This is the constraint. The volume of the box is $V = abc$. Thus we have

$$L(a, b, c) = abc + \lambda(2b + 4a + 6c - 360)$$

where λ is the Lagrange multiplier. Therefore

$$\frac{\partial L}{\partial a} = bc + 4\lambda = 0 \tag{2a}$$

$$\frac{\partial L}{\partial b} = ac + 2\lambda = 0 \tag{2b}$$

$$\frac{\partial L}{\partial c} = ab + 6\lambda = 0 \tag{2c}$$

From (2a) through (2c) we obtain

$$2a^* = b^*, \qquad b^* = 3c^*, \qquad 2a^* = 3c^*.$$

Inserting these equations into the constraint (1) yields

$$a^* = 30 \text{ cm}, \qquad b^* = 60 \text{ cm}, \qquad c^* = 20 \text{ cm}.$$

This is obviously a global maximum.

Problem 3. Let
$$f(\mathbf{x}) = x_1 x_2 \cdots x_n$$
and
$$M := \{ (x_1, x_2, \ldots, x_n) : x_1^2 + x_2^2 + \cdots + x_n^2 = 1 \}.$$
Find the extrema of $f|_M$.

Solution 3. We define
$$r^2 := x_1^2 + x_2^2 + \cdots + x_n^2.$$

From the system

$$\frac{\partial}{\partial x_k}(f(\mathbf{x}) + \lambda(1 - r^2)) = 0, \qquad k = 1, 2, \ldots, n$$

we find

$$2\lambda x_k^{*2} = x_1^* x_2^* \cdots x_n^*, \qquad k = 1, 2, \ldots, n \tag{1}$$

and

$$2\lambda = nx_1^* x_2^* \cdots x_n^* \tag{2}$$

where we have taken into account the constraint $r^2 = 1$. From (1) and (2) we obtain

$$x_k^{*2} = \frac{1}{n}, \qquad k = 1, 2, \ldots, n.$$

Thus at the extrema we have

$$f(\mathbf{x}^*) = \pm\sqrt{\frac{1}{n^n}}.$$

Problem 4.   Show that the angles α, β, γ of a triangle maximize the function

$$f(\alpha, \beta, \gamma) = \sin\alpha \sin\beta \sin\gamma$$

if and only if the triangle is equilateral.

Solution 4.   The constraint is $\alpha + \beta + \gamma = \pi$. Let λ be the Lagrange multiplier. The Lagrangian is

$$L(\alpha, \beta, \gamma) = f(\alpha, \beta, \gamma) + \lambda(\alpha + \beta + \gamma - \pi).$$

Then

$$\frac{\partial L}{\partial \alpha} = 0 \Rightarrow \cos\alpha^* \sin\beta^* \sin\gamma^* + \lambda = 0$$

$$\frac{\partial L}{\partial \beta} = 0 \Rightarrow \sin\alpha^* \cos\beta^* \sin\gamma^* + \lambda = 0$$

$$\frac{\partial L}{\partial \gamma} = 0 \Rightarrow \sin\alpha^* \sin\beta^* \cos\gamma^* + \lambda = 0.$$

We can assume (why?) that $\sin\alpha^* \neq 0$, $\sin\beta^* \neq 0$, $\sin\gamma^* \neq 0$. Thus

$$\cot\alpha^* = \cot\beta^*, \qquad \cot\alpha^* = \cot\gamma^*$$

with the constraint $0 < \alpha^*, \beta^*, \gamma^* < \pi$. We obtain $\alpha^* = \beta^* = \gamma^*$ and therefore $\alpha^* = \beta^* = \gamma^* = \pi/3$. Thus

$$f(\pi/3, \pi/3, \pi/3) = \frac{1}{8}3\sqrt{3}.$$

This is obviously a (global) maximum.

Problem 5.   Consider two smooth non-intersecting curves

$$f(x, y) = 0, \qquad g(x, y) = 0$$

in the plane $\mathbb{R}^2$. Find the necessary conditions for the shortest distance between the two curves. The Lagrange multiplier method must be used. Apply the equations to the curves (circles)

$$f(x, y) = x^2 + y^2 - 1 = 0, \qquad g(x, y) = (x - 3)^2 + (y - 3)^2 - 1 = 0.$$

Solution 5. We denote the first curve with the index 1 and the second with the index 2. Thus the square of the Euclidean distance between the two curves is

$$d^2 = (x_1 - x_2)^2 + (y_1 - y_2)^2$$

with the constraints $f(x_1, y_1) = 0$ and $g(x_2, y_2) = 0$. Thus the Lagrangian is

$$L(x_1, y_1, x_2, y_2) = d^2 + \lambda_1(x_1^2 + y_1^2 - 1) + \lambda_2((x_2 - 3)^2 + (y_2 - 3)^2 - 1).$$

Therefore we have to solve the system of equations

$$\frac{\partial L}{\partial x_1} = 2(x_1 - x_2) + 2\lambda_1 x_1 = 0$$

$$\frac{\partial L}{\partial y_1} = 2(y_1 - y_2) + 2\lambda_1 y_1 = 0$$

$$\frac{\partial L}{\partial x_2} = -2(x_1 - x_2) + 2\lambda_2(x_2 - 3) = 0$$

$$\frac{\partial L}{\partial y_2} = -2(y_1 - y_2) + 2\lambda_2(y_2 - 3) = 0$$

together with the constraints

$$x_1^2 + y_1^2 = 1, \qquad (x_2 - 3)^2 + (y_2 - 3)^2 = 1.$$

Thus we have six equations with six unknowns x_1, y_1, x_2, y_2, λ_1, λ_2. The six equations can be written as

$$x_1(1 + \lambda_1) = x_2$$
$$y_1(1 + \lambda_1) = y_2$$
$$x_2(1 + \lambda_2) = x_1 + 3\lambda_2$$
$$y_2(1 + \lambda_2) = y_1 + 3\lambda_2$$
$$x_1^2 + y_1^2 = 1$$
$$(x_2 - 3)^2 + (y_2 - 3)^2 = 1.$$

We use the first two equations to eliminate x_2 and y_2. We arrive at the four equations

$$x_1(1 + \lambda_1)(1 + \lambda_2) = x_1 + 3\lambda_2$$

$$y_1(1 + \lambda_1)(1 + \lambda_2) = y_1 + 3\lambda_2$$
$$x_1^2 + y_1^2 = 1$$
$$(x_1(1 + \lambda_1) - 3)^2 + (y_1(1 + \lambda_1) - 3)^2 = 1$$

with the four unknowns x_1, y_1, λ_1, λ_2. We consider two cases. First $(1 + \lambda_1)(1 + \lambda_2) = 1$ yields $\lambda_1 = \lambda_2 = 0$, $x_1 = x_2$, $y_1 = y_2$ which cannot satisfy the constraints. Second $(1+\lambda_1)(1+\lambda_2) \neq 1$ yields $x_1 = y_1 = \pm 1/\sqrt{2}$, $x_2 = y_2 = 3 \pm 1/\sqrt{2}$. The shortest distance over the four values is

$$d = \sqrt{18} - 2 = 3\sqrt{2} - 2$$

which is obviously the shortest distance between the two circles.

Problem 6.   Show that the Lagrange multiplier method fails for the following problem. Maximize

$$f(x, y) = -y$$

subject to the constraint

$$g(x, y) = y^3 - x^2.$$

Solution 6.   From

$$L(x, y, \lambda) = f(x, y) + \lambda g(x, y)$$

we obtain the system of equations

$$-2\lambda x = 0, \qquad -1 + 3\lambda y^2 = 0, \qquad -x^2 + y^3 = 0.$$

We have to do a case study. For the first equation to be satisfied, we must have $x = 0$ or $\lambda = 0$. If $\lambda = 0$, the second equation is not satisfied. If $x = 0$, the third equation implies $y = 0$, and once again the second equation cannot be satified.

Using differential forms we find the right solution. We have

$$df \wedge dg = -dy \wedge (3y^2 dy - 2x dx) = -2x dx \wedge dy.$$

Here $\wedge$ is the *wedge product* (also called *exterior product*) which is linear and has the property $dx \wedge dy = -dy \wedge dx$. From $df \wedge dg = 0$ we obtain $x = 0$. Therefore from $y^3 - x^2 = 0$ we obtain $y = 0$ and the maximum $f(0, 0) = 0$.

Chapter 12

Linear Difference Equations

The superposition principle applies to linear difference equations. If two solutions of the linear difference equation are given then the sum of the two solutions is also a solution of the linear difference equation. A solution can also be multiplied with a constant to be a solution again.

Problem 1. Let $n, m \in \mathbb{N}$ with $n > m$. Show that

$$\sum_{k=m}^{n} u_k(v_{k+1} - v_k) \equiv u_k v_k \big|_m^{n+1} - \sum_{k=m}^{n} v_{k+1}(u_{k+1} - u_k) \tag{1}$$

where

$$u_k v_k \big|_m^{n+1} = u_{n+1} v_{n+1} - u_m v_m.$$

Equation (1) is called the formula for *summation by parts*.

Solution 1. We start from the identity

$$u_{k+1} v_{k+1} - u_k v_k \equiv u_k(v_{k+1} - v_k) + v_{k+1}(u_{k+1} - u_k).$$

Therefore

$$\sum_{k=m}^{n} (u_{k+1} v_{k+1} - u_k v_k) \equiv \sum_{k=m}^{n} u_k(v_{k+1} - v_k) + \sum_{k=m}^{n} v_{k+1}(u_{k+1} - u_k).$$

Obviously, the left-hand side is given by

$$\sum_{k=m}^{n} (u_{k+1}v_{k+1} - u_k v_k) \equiv u_{n+1}v_{n+1} - u_m v_m.$$

Problem 2.   Give the solution of the linear difference equation

$$x_{t+1} = 2x_t \tag{1}$$

where $t = 0, 1, 2, \cdots$ and $x_0 = 1$.

Solution 2.   Since (1) is a linear difference equation with constant coefficients we can solve the equation with the ansatz

$$x_t = ar^t$$

where a is a constant. Inserting this ansatz into (1) gives $r = 2$. Therefore the solution of (1) is given by $x_t = a2^t$. Inserting the initial condition $x_0 = 1$ leads to $a = 1$ and therefore

$$x_t = 2^t.$$

Problem 3.   Prove that the set of n (different) elements has exactly 2^n (different) subsets.

Solution 3.   For each n, let X_n denote the number of (different) subsets of a set with n (different) elements. Let S be a set with $n+1$ elements, and designate one of its elements by s. There is a one-to-one correspondence between those subsets of S which do not contain s and those subsets that do contain s (namely, a subset T of the former type corresponds to $T \cup \{s\}$). The former types are all subsets of $S \setminus \{s\}$, a set with n elements, and therefore, it must be the case that

$$X_{n+1} = 2X_n$$

where $X_0 = 1$. Using the result from the previous problem we obtain $X_n = 2^n$.

Problem 4.   (i) Find the solution of the linear difference equation with constant coefficients

$$x_{t+2} - x_{t+1} - x_t = 0 \tag{1}$$

where $t = 0, 1, 2 \ldots$ and $x_0 = 0$, $x_1 = 1$. The numbers x_t are called *Fibonacci numbers*.

(ii) Find

$$g = \lim_{t \to \infty} \frac{x_{t+1}}{x_t}. \tag{2}$$

Solution 4. (i) Since (1) is a linear difference equation with constant coefficients we can solve (1) with the ansatz

$$x_t = a r^t \tag{3}$$

where a is a constant. Inserting ansatz (3) into (1) yields the quadratic equation $r^2 - r - 1 = 0$. The solution to this quadratic equation is given by

$$r_{1,2} = \frac{1}{2} \pm \sqrt{\frac{5}{4}} = \frac{1}{2} \left(1 \pm \sqrt{5} \right).$$

Consequently, the solution to (1) is given by

$$x_t = a_1 r_1^t + a_2 r_2^t$$

where a_1 and a_2 are the two constants of "integration". Imposing the initial condition $x_0 = 0$, $x_1 = 1$ yields

$$x_t = \frac{1}{\sqrt{5}} \left[\left(\frac{1}{2}(1 + \sqrt{5}) \right)^t - \left(\frac{1}{2}(1 - \sqrt{5}) \right)^t \right].$$

(ii) From (2) we find

$$g = \lim_{t \to \infty} \frac{x_{t+1}}{x_t}$$

$$= \lim_{t \to \infty} \frac{x_{t+2}}{x_{t+1}}$$

$$= \lim_{t \to \infty} \frac{x_{t+1} + x_t}{x_{t+1}}$$

$$= 1 + \lim_{t \to \infty} \frac{x_t}{x_{t+1}}$$

$$= 1 + \lim_{t \to \infty} \frac{1}{\frac{x_{t+1}}{x_t}}.$$

Thus g is the solution of the equation $g = 1 + 1/g$ with $g > 1$.

Problem 5. A coin is tossed n times. What is the probability that two heads will turn up in succession somewhere in the sequence of throws?

Solution 5. Let P_n denote the probability that two consecutive heads do not appear in n throws. Obviously

$$P_1 = 1, \qquad P_2 = \frac{3}{4}.$$

If $n > 2$, there are two cases. If the first throw is tails, then two consecutive heads will not appear in the remaining $n - 1$ tosses with probability P_{n-1} (by our choice of notation). If the first throw is heads, the second toss must be tails to avoid two consecutive heads, and then two consecutive heads will not appear in the remaining $n - 2$ throws with probability P_{n-2}. Thus, we obtain the linear difference equation

$$P_n = \frac{1}{2}P_{n-1} + \frac{1}{4}P_{n-2}, \qquad n > 2.$$

This difference equation can be transformed to a more familiar form by multiplying each side by 2^n

$$2^n P_n = 2^{n-1} P_{n-1} + 2^{n-2} P_{n-2}.$$

Defining

$$S_n := 2^n P_n$$

we obtain

$$S_n = S_{n-1} + S_{n-2}.$$

This is the linear difference equation for the Fibonacci sequence. Note that $S_n = F_{n+2}$. Thus, the probability is given by

$$Q_n = 1 - P_n = 1 - \frac{F_{n+2}}{2^n}, \qquad n \geq 1.$$

Problem 6.   (i) Let A be an $n \times n$ matrix with constant entries. Then

$$\mathbf{x}_{t+1} = A\mathbf{x}_t \tag{1}$$

is a system of n linear difference equations with constant coefficients, where $t = 0, 1, 2, \ldots$. Show that the solution of the initial-value problem is given by

$$\mathbf{x}_t = A^t \mathbf{x}_0 \tag{2}$$

where $\mathbf{x}_0$ is the initial column vector in $\mathbb{R}^n$.
(ii) Let

$$A = \begin{pmatrix} 1 & 0 & 0 & 0 \\ 1 & 1 & 0 & 0 \\ 0 & 1 & 1 & 0 \\ -1 & -1 & 0 & 1 \end{pmatrix}.$$

Find the solution of the difference equation (1).

Solution 6.   (i) From (2) it follows that

$$\mathbf{x}_{t+1} = A^{t+1}\mathbf{x}_0 = A^t A\mathbf{x}_0 = AA^t\mathbf{x}_0 = A\mathbf{x}_t.$$

(ii) We set

$$A = I_4 + B$$

where I_4 is the 4×4 unit matrix. Therefore

$$B = \begin{pmatrix} 0 & 0 & 0 & 0 \\ 1 & 0 & 0 & 0 \\ 0 & 1 & 0 & 0 \\ -1 & -1 & 0 & 0 \end{pmatrix}.$$

Now

$$A^t = (I_4 + B)^t = I_4^t + t I_4^{t-1} B + \frac{t(t-1)}{2} I_4^{t-2} B^2 + \cdots + B^t$$

since $[I_4, B] = 0$. Thus

$$A^t = I_4 + tB + \frac{t(t-1)}{2} B^2 + \cdots + B^t.$$

Since

$$B^2 = \begin{pmatrix} 0 & 0 & 0 & 0 \\ 0 & 0 & 0 & 0 \\ 1 & 0 & 0 & 0 \\ -1 & 0 & 0 & 0 \end{pmatrix}$$

and B^3 is the 4×4 zero matrix we obtain

$$A^t = I_4 + tB + \frac{t(t-1)}{2} B^2.$$

Therefore the solution of the initial value problem is given by

$$\mathbf{x}_t = \left(I_4 + tB + \frac{t(t-1)}{2} B^2 \right) \mathbf{x}_0.$$

Problem 7. Sum the finite series

$$S := a_0 + a_1 + \cdots + a_T$$

where $a_0 = 2$ and $a_1 = 5$ for

$$a_{t+2} = 5a_{t+1} - 6a_t \tag{1}$$

with $t = 0, 1, 2, \ldots$.

Solution 7. The first few terms of the a_t-sequence are

$$2, \ 5, \ 13, \ 35, \ 97, \ 275, \ 793, \ldots.$$

A general formula for the nth term is not apparent. We apply the technique of *generating functions.* Let

$$F(x) := a_0 + a_1 x + a_2 x^2 + \cdots + a_t x^t + \cdots \tag{2}$$

be the ansatz for the generating function. We obtain from (2)

$$-5xF(x) = -5a_0 x - 5a_1 x^2 - \cdots - 5a_{t+1} x^{t+2} - \cdots$$
$$6x^2 F(x) = 6a_0 x^2 + 6a_1 x^3 + \cdots + 6a_t x^{t+2} + \cdots .$$

Adding (2) and these equation and using (2), we obtain

$$(1 - 5x + 6x^2)F(x) = a_0 + (a_1 - 5a_0)x$$

so that

$$F(x) = \frac{2 - 5x}{(1 - 2x)(1 - 3x)}.$$

We write this as a sum of partial fractions

$$F(x) = \frac{1}{1 - 2x} + \frac{1}{1 - 3x}.$$

Making use of the geometric series we have

$$F(x) = \sum_{t=0}^{\infty} (2x)^t + \sum_{t=0}^{\infty} (3x)^t \equiv \sum_{t=0}^{\infty} (2^t + 3^t) x^t.$$

Thus

$$a_t = 2^t + 3^t$$

for $t = 0, 1, 2, \dots$. Consequently, the sum is given by

$$S = a_0 + a_1 + \cdots + a_T = \sum_{t=0}^{T} (2^t + 3^t) = \sum_{t=0}^{T} 2^t + \sum_{t=0}^{T} 3^t.$$

Thus

$$S = \frac{2^{T+1} - 1}{2 - 1} + \frac{3^{T+1} - 1}{3 - 1} = 2^{T+1} - 1 + \frac{3^{T+1} - 1}{2} = \frac{2^{T+2} + 3^{T+1} - 3}{2}.$$

The problem can also be solved by solving the linear difference equation with constant coefficients (1) using the ansatz $a_t \propto r^t$ and the initial conditions $a_0 = 2$ and $a_1 = 5$.

Problem 8. An electrical network has the form shown in the following figure, where R_1 and R_2 are given constant resistances. V is the voltage supplied by a source.

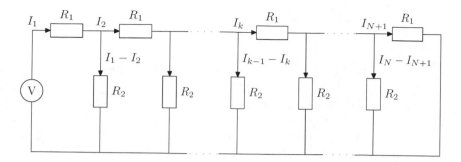

(i) Write down *Kirchhoff's law* for the first loop, the k-th loop and the $(N+1)$-th loop.

(ii) Give the general solution for the difference equation of the k-th loop.

(iii) Determine the constants in the general solution of the difference equation with the boundary conditions (first and $(N+1)$-th loop).

Solution 8. (i) For the first loop which includes the applied voltage Kirchhoff's law provides us with

$$I_1 R_1 + (I_1 - I_2)R_2 - V = 0. \tag{1}$$

For the k-th loop we find

$$I_k R_1 + (I_k - I_{k+1})R_2 - (I_{k-1} - I_k)R_2 = 0. \tag{2}$$

Kirchhoff's law for the last loop $((N+1)$-th loop) gives

$$I_{N+1} R_1 - (I_N - I_{N+1})R_2 = 0. \tag{3}$$

Equation (2) can be written as the difference equation

$$I_{k+1} - \left(2 + \frac{R_1}{R_2}\right) I_k + I_{k-1} = 0 \tag{4}$$

where $k = 1, 2, \ldots, N$.

(ii) Equation (4) is a linear difference equation with constant coefficients. It can therefore be solved with the ansatz

$$I_k = a r^k. \tag{5}$$

The first and last loop provide the boundary conditions. Inserting the ansatz (5) into (4) leads to the algebraic equation

$$r^2 - \left(2 + \frac{R_1}{R_2}\right) r + 1 = 0.$$

The solution of this quadratic equation is given by

$$r_{1,2} = 1 + \frac{R_1}{2R_2} \pm \sqrt{\frac{R_1}{R_2} + \frac{R_1^2}{4R_2^2}}.$$

Consequently, the general solution to the difference equation (4) is given by

$$I_k = ar_1^k + br_2^k$$

where a and b are two constants.

(iii) We impose the boundary conditions (1) and (3). From (1) we find

$$I_2 = \frac{I_1(R_1 + R_2) - V}{R_2}$$

and from (3) we obtain

$$I_N = \frac{(R_1 + R_2)I_{N+1}}{R_2}.$$

If we set $k = 2$ and $k = N$ in these equations we arrive at

$$I_2 = ar_1^2 + br_2^2, \qquad I_N = ar_1^N + br_2^N.$$

It follows that

$$a = \frac{I_2 r_2^N - I_N r_2^2}{r_1^2 r_2^N - r_2^2 r_1^N}, \qquad b = \frac{I_N r_1^2 - I_2 r_1^N}{r_1^2 r_2^N - r_2^2 r_1^N}.$$

Therefore

$$I_k = \left(\frac{I_2 r_2^N - I_N r_2^2}{r_1^2 r_2^N - r_2^2 r_1^N} \right) r_1^k + \left(\frac{I_N r_1^2 - I_2 r_1^N}{r_1^2 r_2^N - r_2^2 r_1^N} \right) r_2^k$$

where r_1, r_2 are given above.

Problem 9.   Consider the coupled system of linear difference equations

$$W_{r+1}(k,l,1) = W_r(k-1,l,1) + e^{-i\pi/4}W_r(k,l-1,2) + e^{i\pi/4}W_r(k,l+1,4)$$
$$(1a)$$

$$W_{r+1}(k,l,2) = e^{i\pi/4}W_r(k-1,l,1) + W_r(k,l-1,2) + e^{-i\pi/4}W_r(k+1,l,3)$$
$$(1b)$$

$$W_{r+1}(k,l,3) = e^{i\pi/4}W_r(k,l-1,2) + W_r(k+1,l,3) + e^{-i\pi/4}W_r(k,l+1,4)$$
$$(1c)$$

$$W_{r+1}(k,l,4) = e^{-i\pi/4}W_r(k-1,l,1) + e^{i\pi/4}W_r(k+1,l,3) + W_r(k,l+1,4)$$
$$(1d)$$

where $k, l \in \{0, 1, 2, \ldots, L-1\}$ and $L \equiv 0$ (periodic boundary conditions). Equations (1a) through (1d) can be written in the matrix form

$$W_{r+1}(k, l, \nu) = \sum_{k', l', \nu'} M(kl\nu, k'l'\nu') W_r(k', l', \nu') \tag{2}$$

where $\nu, \nu' = 1, 2, 3, 4$. Diagonalize M with respect to k and l by applying the discrete Fourier transform.

Solution 9. The *discrete Fourier transform* in two dimensions is given by

$$\hat{W}_r(p, q, \nu) = \sum_{k=0}^{L-1} \sum_{l=0}^{L-1} e^{-2\pi i(pk+ql)/L} W_r(k, l, \nu).$$

The inverse discrete Fourier transformation follows as

$$W_r(k, l, \nu) = \frac{1}{L^2} \sum_{p=0}^{L-1} \sum_{q=0}^{L-1} e^{2\pi i(pk+ql)/L} \hat{W}_r(p, q, \nu).$$

Taking Fourier components on both sides of (1) we find that each equation contains only $\hat{W}_r(p, q, \nu)$ with the same p, q. Consequently, the matrix M is diagonal with respect to p and q. For given p, q we find the 4×4 matrix

$$M(pq\nu|pq\nu') = \begin{pmatrix} \beta^{-p} & \alpha^{-1}\beta^{-q} & 0 & \alpha\beta^q \\ \alpha\beta^{-p} & \beta^{-q} & \alpha^{-1}\beta^p & 0 \\ 0 & \alpha\beta^{-q} & \beta^p & \alpha^{-1}\beta^q \\ \alpha^{-1}\beta^{-p} & 0 & \alpha\beta^p & \beta^q \end{pmatrix}$$

where $\alpha := e^{i\pi/4}$ and $\beta := e^{2\pi i/L}$.

Problem 10. The *Stirling number* of the second kind $S(n, m)$ is the number of partitions of a set with n elements into m classes.
(i) Let

$$X := \{a, b, c, d\}.$$

Thus $n = 4$. Let $m = 2$. Find the partitions.
(ii) Show that $S(n, m)$ satisfies the linear difference equation

$$S(n+1, m) = S(n, m-1) + mS(n, m) \tag{1}$$

with the initial conditions $S(n, 1) = S(n, n) = 1$.

Solution 10. (i) The partitions are

$$\{\,\{\{a\}, \{b, c, d\}\,\}; \,\{\{b\}, \{a, c, d\}\,\}; \,\{\{c\}, \{a, b, d\}\,\}; \,\{\{d\}, \{a, b, c\}\,\};$$

$$\{\{a,b\},\{c,d\}\};\ \{\{a,c\},\{b,d\}\};\ \{\{a,d\},\{b,c\}\}\}\,.$$

Thus we have seven partitions.

(ii) Consider a set with n elements, i.e.

$$X := \{x_1, x_2, \ldots, x_n\}\,.$$

Let $S(n, m-1)$ be the number of partitions into $m-1$ classes. One can obtain $S(n, m-1)$ partitions into m classes of a set with $n+1$ elements $x_1, \ldots, x_{n+1}$ by adding to each partition a new class consisting of only the element x_{n+1}. The element x_{n+1} can be added to each of the already existing m classes of a partition of X in m distinct ways. These two procedures yield, without repetitions, all the partitions of the set X into m classes. Equation (1) follows. The solution to (1), taking into account the initial conditions, is given by

$$S(n, m) = \frac{1}{m!} \sum_{k=0}^{m-1} (-1)^k \binom{m}{k} (m-k)^n\,.$$

Problem 11.   Solve the linear recurrence relation

$$x_{n+1} = 1 + \sum_{j=0}^{n-1} x_j, \qquad x_0 = 1\,.$$

Solution 11.   We have $x_1 = 1$, since the empty sum is zero. For $n \geq 1$, we have

$$x_{n+1} = 1 + \sum_{j=0}^{n-1} x_j = \left(1 + \sum_{j=0}^{n-2} x_j\right) + x_{n-1} = x_n + x_{n-1}$$

and so by induction x_n is the Fibonacci number for n.

Problem 12.   Let n be a positive integer and $f(j) = j(j-1)(j-2)$ with $j = 1, 2, \ldots, n$. Let $a_j := f(j+1) - f(j)$. By calculating $\sum_{j=1}^{n} a_j$ show that

$$\sum_{j=1}^{n} j^2 = \frac{n(n+1)(2n+1)}{6}\,.$$

Solution 12.   From $a_j := f(j+1) - f(j)$ we obtain

$$\sum_{j=1}^{n} a_j = f(n+1) - f(1) = f(n+1) = (n+1)n(n-1)\,.$$

Now

$$a_j = f(j+1) - f(j)$$
$$= (j+1)j(j-1) - j(j-1)(j-2)$$
$$= 3j^2 - 3j .$$

If follows that

$$\sum_{j=1}^{n} a_j = 3\sum_{j=1}^{n} j^2 - 3\sum_{j=1}^{n} j = (n+1)n(n-1) .$$

Since $\sum_{j=1}^{n} j = n(n+1)/2$ we obtain

$$\sum_{j=1}^{n} j^2 = \frac{(n+1)n(n-1)}{3} + \frac{n(n+1)}{2} = \frac{(n+1)n(2n+1)}{6} .$$

Problem 13.   The *unilateral z-transform* is defined by

$$\mathcal{Z}(z) := \sum_{n=0}^{\infty} x(n)z^{-n}$$

i.e. the input starts at $n = 0$. The inverse z-transform is given by

$$x(n) = \frac{1}{2\pi i} \oint \mathcal{Z}(z)z^{n-1}dz$$

where $\oint$ denotes integration around a circular path in the complex plane centered at the origin. Apply the *z-transform* to find the solution of the linear third order difference equation

$$x_{n+3} - 3x_{n+2} + 3x_{n+1} - x_n = 0$$

with the initial conditions $x_0 = 0$, $x_1 = 1$, $x_2 = 4$.

Solution 13.   We have

$$\mathcal{Z}(x_{n+\ell})(z) = z^{\ell}\mathcal{Z}(x_n)(z) - \sum_{\nu=0}^{\ell-1} v^2 z^{\ell-\nu}, \quad \ell = 1,2,3 .$$

Thus

$$z^3\mathcal{Z}(x_n)(z) - z^2 - 4z - 3(z^2\mathcal{Z}(x_n)(z) - z) + 3z\mathcal{Z}(x_n)(z) - \mathcal{Z}(x_n)(z) = 0 .$$

Therefore

$$\mathcal{Z}(x_n)(z) = \frac{z(z+1)}{(z-1)^3}, \quad |z| > 1 .$$

Since

$$\frac{z(z+1)}{(z-1)^3} \equiv \frac{1}{z-1} + \frac{3}{(z-1)^2} + \frac{2}{(z-1)^3}$$

we obtain

$$x_n = \binom{n-1}{0} + 3\binom{n-1}{1} + 2\binom{n-1}{2}.$$

For $n > 3$ we have

$$x_n = 1 + 3(n-1) + (n-1)(n-2) = n^2.$$

Problem 14.   Consider a function $f(j)$, where j is integer-valued. Find multiplication rules of the form

$$(f * g)(j) = \sum_{k,\ell} c_{jk\ell} f(k) g(\ell)$$

and difference operators Δ of the form

$$\Delta f(j) = \sum_k d_{jk} f(k)$$

where the $c_{jk\ell}$ and the d_{jk} are real constants such that the following properties hold:
(i) * is commutative and associative
(ii) if f is constant, then $f * g = fg$ (normal multiplication)
(iii) if f is constant, then $\Delta f = 0$
(iv) the *Leibniz rule*

$$\Delta(f * g) = f * (\Delta g) + (\Delta f) * g$$

holds
(v) the function f is periodic, i.e. there is a fixed positive N such that $f(j + N) = f(j)$.

Solution 14.   A solution which satisfies these conditions is

$$(f * g)(j) = \langle f \rangle g(j) + \langle g \rangle f(j) - \langle f \rangle \langle g \rangle$$

where $\langle f \rangle$ is the average

$$\langle f \rangle := \frac{1}{N} \sum_{j=1}^{N} f(j).$$

The difference operator Δ is the *forward difference*

$$\Delta f(j) = f(j+1) - f(j).$$

Chapter 13

Linear Differential Equations

The superposition principle applies to linear differential equations. If two solutions of the linear differential equation are given then the sum of the two solutions is also a solution of the linear differential equation. A solution can also be multiplied with a constant to be a solution again.

Problem 1. Find the general solution for the *driven damped harmonic oscillator*

$$\frac{d^2u}{dt^2} + \alpha\frac{du}{dt} + \omega^2 u = k_1\cos(\Omega_1 t) + k_2\cos(\Omega_2 t) \tag{1}$$

where $\alpha > 0$ and $\Omega_1 \neq \Omega_2$. Discuss the dependence on Ω_1 and Ω_2.

Solution 1. The general solution of equation (1) is given by

$$u = u_h + u_p$$

where u_h is the general solution to the homogeneous equation and u_p is a particular solution to the inhomogeneous equation. We first determine the general solution of the homogeneous differential equation

$$\frac{d^2u}{dt^2} + \alpha\frac{du}{dt} + \omega^2 u = 0. \tag{2}$$

Since (2) is a linear differential equation with constant coefficients we can solve the equation with the exponential ansatz

$$u(t) \propto e^{\lambda t}.$$

133

Inserting this ansatz into (2) yields the quadratic equation

$$\lambda^2 + \alpha\lambda + \omega^2 = 0.$$

The roots are

$$\lambda_\pm = -\frac{\alpha}{2} \pm \sqrt{\left(\frac{\alpha}{2}\right)^2 - \omega^2}.$$

Consequently, the general solution of the homogeneous part of the linear differential equation is given by one of the following cases

$$\omega^2 = \left(\frac{\alpha}{2}\right)^2 : \quad u_h(t) = (C + Dt)e^{-\alpha t/2}$$

$$\omega^2 < \left(\frac{\alpha}{2}\right)^2 : \quad u_h(t) = Ce^{(-\alpha/2+\sqrt{(\alpha/2)^2-\omega^2})t} + De^{(-\alpha/2-\sqrt{(\alpha/2)^2-\omega^2})t}$$

$$\omega^2 > \left(\frac{\alpha}{2}\right)^2 : \quad u_h(t) = (C\cos\sqrt{\omega^2-(\alpha/2)^2}t + D\sin\sqrt{\omega^2-(\alpha/2)^2}t)e^{-\alpha t/2}$$

where C and D are the constants of integration. To find the particular solution to (1) we make the ansatz

$$u_p(t) = A_1\cos(\Omega_1 t) + B_1\sin(\Omega_1 t) + A_2\cos(\Omega_2 t) + B_2\sin(\Omega_2 t).$$

We find the following linear equations for A_1, A_2, B_1 and B_2

$$k_1 = -A_1\Omega_1^2 + \alpha B_1\Omega_1 + \omega^2 A_1$$
$$0 = -B_1\Omega_1^2 - \alpha A_1\Omega_1 + \omega^2 B_1$$
$$k_2 = -A_2\Omega_2^2 + \alpha B_2\Omega_2 + \omega^2 A_2$$
$$0 = -B_2\Omega_2^2 - \alpha A_2\Omega_2 + \omega^2 B_2.$$

Consequently, we find that the general solution to (1) is given by

$$u(t) = u_h(t) + \sum_{j=1}^{2} \frac{k_j}{(\omega^2 - \Omega_j^2)^2 + \alpha^2\Omega_j^2}[(\omega^2 - \Omega_j^2)\cos(\Omega_j t) + \alpha\Omega_j\sin(\Omega_j t)].$$

Problem 2.   The *motion of a charge q* in an electromagnetic field is given by

$$m\frac{d\mathbf{v}}{dt} = q(\mathbf{E} + \mathbf{v} \times \mathbf{B}) \tag{1}$$

where m denotes the mass and $\mathbf{v}$ the velocity. Assume that

$$\mathbf{E} = \begin{pmatrix} E_1 \\ E_2 \\ E_3 \end{pmatrix}, \qquad \mathbf{B} = \begin{pmatrix} B_1 \\ B_2 \\ B_3 \end{pmatrix}$$

are constant fields. Find the solution of the initial value problem.

Solution 2. Equation (1) can be written in the form

$$
\begin{pmatrix} dv_1/dt \\ dv_2/dt \\ dv_3/dt \end{pmatrix} = \frac{q}{m} \begin{pmatrix} 0 & B_3 & -B_2 \\ -B_3 & 0 & B_1 \\ B_2 & -B_1 & 0 \end{pmatrix} \begin{pmatrix} v_1 \\ v_2 \\ v_3 \end{pmatrix} + \frac{q}{m} \begin{pmatrix} E_1 \\ E_2 \\ E_3 \end{pmatrix}.
$$

We set

$$
B_j \rightarrow \frac{q}{m} B_j, \qquad E_j \rightarrow \frac{q}{m} E_j.
$$

Thus

$$
\begin{pmatrix} dv_1/dt \\ dv_2/dt \\ dv_3/dt \end{pmatrix} = \begin{pmatrix} 0 & B_3 & -B_2 \\ -B_3 & 0 & B_1 \\ B_2 & -B_1 & 0 \end{pmatrix} \begin{pmatrix} v_1 \\ v_2 \\ v_3 \end{pmatrix} + \begin{pmatrix} E_1 \\ E_2 \\ E_3 \end{pmatrix}. \tag{2}
$$

Equation (2) is a system of nonhomogeneous linear differential equations with constant coefficients. The solution to the homogeneous equation

$$
\begin{pmatrix} dv_1/dt \\ dv_2/dt \\ dv_3/dt \end{pmatrix} = \begin{pmatrix} 0 & B_3 & -B_2 \\ -B_3 & 0 & B_1 \\ B_2 & -B_1 & 0 \end{pmatrix} \begin{pmatrix} v_1 \\ v_2 \\ v_3 \end{pmatrix}
$$

is given by

$$
\begin{pmatrix} v_1(t) \\ v_2(t) \\ v_3(t) \end{pmatrix} = e^{tM} \begin{pmatrix} v_1(0) \\ v_2(0) \\ v_3(0) \end{pmatrix}
$$

where M is the matrix of the right hand side and $v_j(0) = v_j(t = 0)$. The solution of the system of nonhomogeneous linear differential equations can be found with the help of the method called *variation of constants*. One sets

$$
\mathbf{v}(t) = e^{tM} \mathbf{f}(t)
$$

where $\mathbf{f} : \mathbb{R} \rightarrow \mathbb{R}^3$ is some differentiable curve. Then differentiation yields

$$
\frac{d\mathbf{v}}{dt} = M e^{tM} \mathbf{f}(t) + e^{tM} \frac{d\mathbf{f}}{dt}.
$$

Inserting this ansatz into (2) yields

$$
M\mathbf{v}(t) + \mathbf{E} = M e^{tM} \mathbf{f}(t) + e^{tM} \frac{d\mathbf{f}}{dt} = M\mathbf{v}(t) + e^{tM} \frac{d\mathbf{f}}{dt}.
$$

Consequently

$$
\frac{d\mathbf{f}}{dt} = e^{-tM} \mathbf{E}.
$$

By integration we obtain

$$
\mathbf{f}(t) = \int_0^t e^{-sM} \mathbf{E}\, ds + \mathbf{K}
$$

where

$$\mathbf{K} = \begin{pmatrix} K_1 \\ K_2 \\ K_3 \end{pmatrix}.$$

Therefore we obtain the general solution of the initial value problem of the nonhomogeneous system (2), namely

$$\mathbf{v}(t) = e^{tM} \left(\int_0^t e^{-sM} \mathbf{E} ds + \mathbf{K} \right)$$

where

$$\mathbf{K} = \begin{pmatrix} v_1(0) \\ v_2(0) \\ v_3(0) \end{pmatrix}.$$

We now have to calculate e^{tM} and e^{-sM}. We find that

$$M^2 = \begin{pmatrix} -B_2^2 - B_3^2 & B_1 B_2 & B_1 B_3 \\ B_1 B_2 & -B_1^2 - B_3^2 & B_2 B_3 \\ B_1 B_3 & B_2 B_3 & -B_1^2 - B_2^2 \end{pmatrix}$$

and $M^3 = M^2 M = -B^2 M$, where $B^2 = B_1^2 + B_2^2 + B_3^2$ and

$$B := \sqrt{B_1^2 + B_2^2 + B_3^2}.$$

Therefore $M^4 = -B^2 M^2$. Since

$$e^{tM} := \sum_{k=0}^{\infty} \frac{(tM)^k}{k!} = I_3 + \frac{tM}{1!} + \frac{t^2 M^2}{2!} + \frac{t^3 M^3}{3!} + \frac{t^4 M^4}{4!} + \frac{t^5 M^5}{5!} + \cdots$$

where I_3 denotes the 3×3 unit matrix, we obtain

$$e^{tM} = I_3 + M \left(t - \frac{t^3}{3!} B^2 + \frac{t^5}{5!} B^4 - \cdots \right) + M^2 \left(\frac{t^2}{2!} - \frac{t^4}{4!} B^2 + \frac{t^6}{6!} B^4 - \cdots \right).$$

Thus

$$e^{tM} = I_3 + \frac{M}{B} \left(tB - \frac{t^3}{3!} B^3 + \cdots \right) - \frac{M^2}{B^2} \left(1 - 1 - \frac{t^2 B^2}{2!} + \frac{t^4 B^4}{4!} - \cdots \right)$$

$$= I_3 + \frac{M}{B} \sin(Bt) + \frac{M^2}{B^2} (1 - \cos(Bt)).$$

Therefore

$$e^{tM} = I_3 + \frac{M}{B} \sin(Bt) + \frac{M^2}{B^2} (1 - \cos(Bt))$$

and

$$e^{-sM} = I_3 - \frac{M}{B} \sin(Bs) + \frac{M^2}{B^2} (1 - \cos(Bs)).$$

Since

$$\int_0^t e^{-sM} \mathbf{E} ds = \int_0^t \left(I_3 - \frac{M}{B} \sin(Bs) + \frac{M^2}{B^2} (1 - \cos(Bs)) \right) \mathbf{E} ds$$

$$= \mathbf{E}t + \frac{M\mathbf{E}}{B^2} \cos(Bt) - \frac{M\mathbf{E}}{B^2} + \frac{M^2\mathbf{E}t}{B^2} - \frac{M^2\mathbf{E}}{B^3} \sin(Bt)$$

we find as the solution of the initial value problem of system (2)

$$\mathbf{v}(t) = \mathbf{E}t \left(1 + \frac{M^2}{B^2} \right) - \frac{M^2\mathbf{E}}{B^3} \sin(Bt) + \frac{M\mathbf{v}(0)}{B} \sin(Bt)$$

$$+ \frac{M\mathbf{E}}{B^2} (1 - \cos(Bt)) + \frac{M^2\mathbf{v}(0)}{B^2} (1 - \cos(Bt)) + \mathbf{v}(0)$$

where $\mathbf{v}(t = 0) \equiv \mathbf{v}(0)$.

Problem 3. Solve the linear differential equation

$$\frac{d^2u}{dx^2} + u = 0 \tag{1}$$

with the following boundary conditions

$$u(0) = 1, \qquad u(1) = 1 \tag{2a}$$

$$u(0) = 1, \qquad u(\pi) = -1 \tag{2b}$$

$$u(0) = 1, \qquad u(\pi) = -2. \tag{2c}$$

Discuss the result.

Solution 3. The general solution to (1) is

$$u(x) = C_1 \cos x + C_2 \sin x$$

where C_1 and C_2 are the constants of integration. Imposing the boundary condition (2a) yields

$$u(0) = 1, \ u(1) = 1 \Rightarrow C_1 = 1, \quad C_2 = \frac{1 - \cos 1}{\sin 1}.$$

Thus with the boundary condition (2a) we have one solution. Imposing the boundary condition (2b) yields

$$u(0) = 1, \quad u(\pi) = -1 \Rightarrow C_1 = 1, \quad C_2 = \text{arbitrary}.$$

Since C_2 is arbitrary we have infinitely-many solutions for boundary condition (2b). Imposing the boundary condition (2c) gives

$$u(0) = 1, \quad u(\pi) = -2 \Rightarrow C_1 = 1, \quad \underbrace{-2 = 1(-1) + C_2 \cdot 0}_{\text{cannot be satisfied}}.$$

Consequently, no solution exists for boundary condition (2c). *Boundary value problems* can lead to (a) unique solutions, (b) arbitrarily many solutions, (c) no solution.

Problem 4.   Let

$$H(\mathbf{p}, \mathbf{q}) = \frac{1}{2}\left(\frac{p_1^2}{m} + \frac{p_2^2}{m}\right) + \frac{1}{2}k(q_2 - q_1)^2 \tag{1}$$

be a Hamilton function, where k is a positive constant. Solve Hamilton's equations of motion for this Hamilton function. *Hamilton's equations of motion* are given by

$$\frac{dq_j}{dt} = \frac{\partial H}{\partial p_j}, \quad \frac{dp_j}{dt} = -\frac{\partial H}{\partial q_j} \tag{2}$$

where $j = 1, 2$.

Solution 4.   Inserting the Hamilton function (1) into system (2) yields

$$\frac{dq_1}{dt} = \frac{p_1}{m}, \quad \frac{dq_2}{dt} = \frac{p_2}{m} \tag{3a}$$

$$\frac{dp_1}{dt} = k(-q_1 + q_2), \quad \frac{dp_2}{dt} = k(q_1 - q_2). \tag{3b}$$

This is an autonomous system of first-order linear differential equations with constant coefficients. Let

$$A := \begin{pmatrix} 0 & 0 & \dfrac{1}{m} & 0 \\ 0 & 0 & 0 & \dfrac{1}{m} \\ -k & k & 0 & 0 \\ k & -k & 0 & 0 \end{pmatrix}.$$

Then system (3) can be written in the matrix form

$$\begin{pmatrix} dq_1/dt \\ dq_2/dt \\ dp_1/dt \\ dp_2/dt \end{pmatrix} = A \begin{pmatrix} q_1 \\ q_2 \\ p_1 \\ p_2 \end{pmatrix}.$$

Thus the general solution of the initial value problem is given by

$$\begin{pmatrix} q_1(t) \\ q_2(t) \\ p_1(t) \\ p_2(t) \end{pmatrix} = e^{tA} \begin{pmatrix} q_1 \\ q_2 \\ p_1 \\ p_2 \end{pmatrix} \Bigg|_{q_1 \to q_1(0), \dots, p_2 \to p_2(0)}.$$

The exponential function $\exp(tA)$ can be evaluated by determining the eigenvalues and normalized eigenvectors of A. The eigenvalues of A are given by

$$\lambda_{1,2} = 0, \qquad \lambda_{3,4} = \pm i\omega$$

where $\omega^2 = 2k/m$. Then the solution of the initial value problem is given by

$$q_1(t) = \frac{q_{10} - q_{20}}{2}\cos(\omega t) + \frac{p_{10} - p_{20}}{2\omega m}\sin(\omega t) + \frac{p_{10} + p_{20}}{2m}t + \frac{q_{10} + q_{20}}{2}$$

$$q_2(t) = \frac{q_{20} - q_{10}}{2}\cos(\omega t) - \frac{p_{10} - p_{20}}{2\omega m}\sin(\omega t) + \frac{p_{10} + p_{20}}{2m}t + \frac{q_{10} + q_{20}}{2}$$

where $q_{j0} \equiv q_j(t = 0)$ and $p_{j0} \equiv p_j(t = 0)$ are the initial values. The momenta p_1 and p_2 are given by (3a) and (3b), respectively.

Problem 5. (i) Show that the substitution

$$u = -\frac{1}{v}\frac{dv}{dt} \tag{1}$$

reduce the nonlinear differential equation (*Riccati equation*)

$$\frac{du}{dt} = u^2 + t, \qquad u(0) = 1 \tag{2}$$

to the linear second-order differential equation

$$\frac{d^2v}{dt^2} + tv = 0. \tag{3}$$

(ii) Find the inverse transformation of (1).

Solution 5. (i) From (1) we obtain

$$\frac{du}{dt} = \frac{1}{v^2}\left(\frac{dv}{dt}\right)^2 - \frac{1}{v}\frac{d^2v}{dt^2}. \tag{4}$$

Inserting (4) and (1) into (2) yields (3).
(ii) Since

$$u(t) = -\frac{d}{dt}(\ln v)$$

we find

$$v(t) = \exp\left(-\int^t u(s)ds\right).$$

Equation (2) has no solution in terms of the elementary functions. *Bessel functions* are needed to solve it.

Problem 6.   Let f be an analytic function of x and u. In *Picard's method* one approximates a solution of a first order differential equation

$$\frac{du}{dx} = f(x, u)$$

with initial conditions $u(x_0) = u_0$ as follows. Integrating both sides yields

$$u(x) = u_0 + \int_{x_0}^{x} f(s, u(s)) ds .$$

Now starting with u_0 this formula can be used to approach the exact solution iteratively if the series converges. The next approximation is given by

$$u_{k+1}(x) = u_0 + \int_{x_0}^{x} f(s, u_k(s)) ds, \qquad k = 0, 1, 2, \ldots .$$

Apply this approach to the linear differential equation

$$\frac{du}{dx} = x + u$$

where $x_0 = 0$, $u(x_0) = 1$.

Solution 6.   Five steps in Picard's method provide

$$u(x) \approx 1 + x + x^2 + \frac{x^3}{3} + \frac{x^4}{12} + \frac{x^5}{60} + \frac{x^6}{720}$$

What is the radius of convergence of this series?

Problem 7.   Find the solution of the initial value problem

$$\frac{d^3 u}{dt^3} + u = 0$$

with $u(t = 0) = u_0$, $du(t = 0)/dt = u_{t0}$, $d^2 u(t = 0)/dt^2 = u_{tt0}$ and u is a real-valued function.

Solution 7.   From

$$u(t) \propto e^{\lambda t}, \quad du/dt \propto \lambda e^{\lambda t}, \quad d^2 u/dt^2 \propto \lambda^2 e^{\lambda t}, \quad d^3 u/dt^3 \propto \lambda^3 e^{\lambda t}$$

Thus we obtain the characteristic equation $\lambda^3 + 1 = 0$ with the solutions

$$\lambda_1 = -1, \quad \lambda_2 = -e^{i2\pi/3} \equiv \frac{1}{2}(1 - i\sqrt{3}), \quad \lambda_3 = -e^{i4\pi/3} \equiv \frac{1}{2}(1 + i\sqrt{3}) .$$

Thus we have the general solution

$$u(t) = Ae^{-t} + Be^{t/2}e^{-i\sqrt{3}t/2} + Ce^{t/2}e^{i\sqrt{3}t/2}$$

where A, B, C are arbitrary constants. We can also write with

$$B = B_1 + iB_2, \qquad C = C_1 + iC_2$$

$(A, B_1, B_2, C_1, C_2$ real$)$

$$u(t) = Ae^{-t} + ((B_1 + C_1)\cos(\sqrt{3}t/2) + (B_2 - C_2)\sin(\sqrt{3}t/2))e^{t/2}$$
$$+ i((B_2 + C_2)\cos(\sqrt{3}t/2) + (C_1 - B_1)\sin(\sqrt{3}t/2))e^{t/2}.$$

Thus we can set $B_2 + C_2 = 0$ and $C_1 - B_1 = 0$. Thus

$$u(t) = Ae^{-t} + (2B_1\cos(\sqrt{3}t/2) - 2C_2\sin(\sqrt{3}t/2))e^{t/2}.$$

Now we can set $\widetilde{B} = 2B_1$ and $\widetilde{C} = -2C_2$ and arrive at

$$u(t) = Ae^{-t} + \widetilde{B}\cos(\sqrt{3}t/2)e^{t/2} + \widetilde{C}\sin(\sqrt{3}t/2)e^{t/2}.$$

Finally we have to insert the initial conditons. We find

$$u_0 = A + \widetilde{B}$$
$$u_{t0} = -A + \frac{1}{2}B + \frac{\sqrt{3}}{2}\widetilde{C}$$
$$u_{tt0} = A - \frac{1}{2}\widetilde{B} + \frac{\sqrt{3}}{2}\widetilde{C}.$$

Problem 8. Consider the linear differential equation

$$\frac{d\mathbf{x}}{dt} = P(t)\mathbf{x}$$

where $P(t)$ is periodic with principal period T and differentiable. Thus T is the smallest positive number for which $P(t+T) = P(t)$ and $-\infty < t < \infty$. Can we conclude that all solutions are periodic? For example, consider

$$\frac{dx}{dt} = (1 + \sin t)x.$$

Solution 8. Thus $P(t) = 1 + \sin t$ has period 2π, but all solutions are given by

$$x(t) = C\exp(t - \cos t)$$

where C is any constant, so only the solution $x(t) = 0$ is periodic.

Problem 9.   Solve the initial value problem for the system of linear differential equations

$$\frac{dc_0}{dt} = \frac{1}{2}i\Omega e^{-i\phi}e^{i(\omega-\nu)t}c_1$$

$$\frac{dc_1}{dt} = \frac{1}{2}i\Omega e^{i\phi}e^{-i(\omega-\nu)t}c_0$$

where Ω, ω, ν are constant frequencies. Note the the system depends explicitly on the time t. Then study the special case $\omega = \nu$.

Solution 9.   Differentiating the first equation with respect to t and inserting the second equation provides a second order differential equation with constant coefficients

$$\frac{d^2c_0}{dt^2} - i(\omega - \nu)\frac{dc_0}{dt} + \frac{1}{4}\Omega^2 c_0 = 0\,.$$

This equation can be solved with the exponential ansatz. Using

$$\frac{dc_0(0)}{dt} = \frac{1}{2}i\Omega e^{-i\phi}c_1(0)$$

we obtain the solution

$$c_0(t) = e^{i\Delta t/2}\left(\cos(ft/2) - i\frac{\Delta}{f}\sin(ft/2)\right)c_0(0) + i\frac{\Omega}{f}e^{i\Delta t/2}e^{-i\phi}\sin(ft/2)c_1(0)$$

$$c_1(t) = i\frac{\Omega}{f}e^{-i\Delta t/2}e^{i\phi}\sin(ft/2)c_0(0) + e^{-i\Delta t/2}\left(\cos(ft/2) + i\frac{\Delta}{f}\sin(ft/2)\right)c_1(0)$$

where $\Delta := \omega - \nu$ and

$$f := \sqrt{\Omega^2 + (\omega - \nu)^2}\,.$$

For the special case $\omega = \nu$ we have

$$c_0(t) = \cos(ft/2)c_0(0) + ie^{-i\phi}\sin(ft/2)c_1(0)$$
$$c_1(t) = ie^{i\phi}\sin(ft/2)c_0(0) + \cos(ft/2)c_1(0)\,.$$

Problem 10.   The *time-independent Schrödinger equation* (eigenvalue equation) for a one-particle problem is given by

$$\left(-\frac{\hbar^2}{2m}\Delta + U(\mathbf{r})\right)u = Eu \tag{1}$$

where

$$\Delta := \frac{\partial^2}{\partial x_1^2} + \frac{\partial^2}{\partial x_2^2} + \frac{\partial^2}{\partial x_3^2}\,.$$

Show that if the potential energy $U(\mathbf{r})$ can be written as a sum of functions of a single coordinate,

$$U(\mathbf{r}) = U_1(x_1) + U_2(x_2) + U_3(x_3)$$

then the time-independent Schrödinger equation can be decomposed into a set of one-dimensional differential equations of the form

$$\frac{d^2 u_j(x_j)}{dx_j^2} + \frac{2m}{\hbar^2}\left(E_j - U_j(x_j)\right) u_j(x_j) = 0, \qquad j = 1, 2, 3$$

with the help of the *separation ansatz*

$$u(\mathbf{r}) = u_1(x_1)u_2(x_2)u_3(x_3) \tag{2}$$

and $E = E_1 + E_2 + E_3$.

Solution 10. Substituting the separation ansatz (2) into (1) and dividing by $u_1 u_2 u_3$, we obtain

$$\sum_{j=1}^{3}\left(\frac{1}{u_j}\frac{d^2 u_j}{dx_j^2} - \frac{2m}{\hbar^2}U_j(x_j)\right) = -\frac{2m}{\hbar^2}E.$$

Since the terms in each bracket of the sum contain independent variables, the equality can be valid for all (x_1, x_2, x_3) only if each bracket is a constant, i.e. if

$$\frac{1}{u_j}\frac{d^2 u_j}{dx_j^2} - \frac{2m}{\hbar^2}U_j = -\frac{2m}{\hbar^2}E_j, \qquad j = 1, 2, 3$$

where the E_j are constants, and $E = E_1 + E_2 + E_3$.

Problem 11. (i) Show that the general solution of the one-dimensional *wave equation*

$$\frac{1}{c^2}\frac{\partial^2 u}{\partial t^2} = \frac{\partial^2 u}{\partial x^2} \tag{1}$$

is given by

$$u(x, t) = f(x - ct) + g(x + ct) \tag{2}$$

where f and g are smooth functions and c is a positive constant.
(ii) Solve the initial value problem

$$u(t = 0, x) = u_0(x)$$

$$\frac{\partial u}{\partial t}(t = 0, x) = u_1(x).$$

Solution 11.   (i) Let $s := x - ct$ and $r := x + ct$. Then

$$\frac{\partial u}{\partial t} = \frac{df}{ds}\frac{\partial s}{\partial t} + \frac{dg}{dr}\frac{\partial r}{\partial t} = -c\frac{df}{ds} + c\frac{dg}{dr}$$

and

$$\frac{\partial^2 u}{\partial t^2} = c^2\frac{d^2 f}{ds^2} + c^2\frac{d^2 g}{dr^2}. \tag{3}$$

Analogously,

$$\frac{\partial^2 u}{\partial x^2} = \frac{d^2 f}{ds^2} + \frac{d^2 g}{dr^2}. \tag{4}$$

Inserting (3) and (4) into (1) shows that (2) is the (general) solution. Equation (2) is the general solution of the one-dimensional wave equation since f and g are arbitrary.
(ii) From (2) we obtain

$$u_0(x) = f(x) + g(x).$$

Since

$$\frac{\partial u(x,t)}{\partial t} = -c\frac{df(x-ct)}{ds} + c\frac{dg(x+ct)}{dr}$$

it follows that

$$u_1(x) = -c\frac{df(x)}{dx} + c\frac{dg(x)}{dx}.$$

Integrating this equation yields

$$A + \int_a^x u_1(\alpha)d\alpha = -cf(x) + cg(x)$$

where A, a are two arbitrary constants. Using $f(x) = u_0(x) - g(x)$ and $g(x) = u_0(x) - f(x)$ provides the two equations

$$2cf(x) = cu_0(x) - A - \int_a^x u_1(\alpha)d\alpha$$

$$2cg(x) = cu_0(x) + A + \int_a^x u_1(\alpha)d\alpha.$$

With the translation $x \to x - ct$, $x \to x + ct$ it follows that

$$2cf(x - ct) = cu_0(x - ct) - A - \int_a^{x-ct} u_1(\alpha)d\alpha$$

$$2cg(x + ct) = cu_0(x + ct) + A + \int_a^{x+ct} u_1(\alpha)d\alpha.$$

Adding these two equations yields

$$u(x,t) = \frac{1}{2}\left(u_0(x+ct) + u_0(x-ct)\right) + \frac{1}{2c}\int_{x-ct}^{x+ct} u_1(\alpha)d\alpha.$$

Consider the Cauchy problem of the the partial differential equation

$$\frac{\partial^2 u}{\partial t^2} - c^2\frac{\partial^2 u}{\partial x^2} = f(t,x)$$

with

$$u(t,x)|_{t=0} = \phi(x), \qquad \left.\frac{\partial u}{\partial t}\right|_{t=0} = \psi(x)$$

where $\phi \in C^2(\mathbb{R})$, $\psi \in C^1(\mathbb{R})$ and the function f is continuous together with the first derivative with respect to x in the half-plane

$$\{\, t \geq 0, \; -\infty < x < +\infty \,\}.$$

The solution is (d'Alembert's formula)

$$u(x,t) = \frac{1}{2c}\int_0^t \int_{x-c(t-\tau)}^{x+c(t-\tau)} f(\tau,\eta)d\eta d\tau$$
$$+ \frac{1}{2c}\int_{x-ct}^{x+ct} \psi(\eta)d\eta + \frac{1}{2}(\phi(x+ct) + \phi(x-ct)).$$

Problem 12.   The *three-dimensional wave equation* is given by

$$\frac{1}{c^2}\frac{\partial^2 u}{\partial t^2} = \frac{\partial^2 u}{\partial x^2} + \frac{\partial^2 u}{\partial y^2} + \frac{\partial^2 u}{\partial z^2} \equiv \Delta u. \tag{1}$$

(i) Express the wave equation in spherical coordinates. Omit the angle part and find the wave equation for the radial part.
(ii) Find the general solution of the wave equation which only includes the radial part.

Solution 12.   (i) The *spherical coordinates* are given by

$$x(r,\phi,\theta) = r\cos\phi\sin\theta \qquad y(r,\phi,\theta) = r\sin\phi\sin\theta \qquad z(r,\phi,\theta) = r\cos\theta$$

where $0 \leq \phi < 2\pi$, $0 < \theta < \pi$ and $r > 0$. Let

$$u(x(r,\phi,\theta), y(r,\phi,\theta), z(r,\phi,\theta), t) = v(r,\phi,\theta,t).$$

Applying the chain rule we find from (1) that

$$\frac{1}{r^2}\frac{\partial}{\partial r}\left(r^2\frac{\partial}{\partial r}\right)v + \frac{1}{r^2\sin\theta}\frac{\partial}{\partial\theta}\left(\sin\theta\frac{\partial}{\partial\theta}\right)v + \frac{1}{r^2\sin^2\theta}\frac{\partial^2}{\partial\phi^2}v = \frac{1}{c^2}\frac{\partial^2 v}{\partial t^2}.$$

If v is a *spherically-symmetric solution* of the wave equation, i.e. $\partial v/\partial \theta = 0$ and $\partial v/\partial \phi = 0$ the wave equation becomes

$$\frac{1}{r^2}\frac{\partial}{\partial r}\left(r^2\frac{\partial v}{\partial r}\right) = \frac{1}{c^2}\frac{\partial^2 v}{\partial t^2}$$

or equivalently

$$\frac{1}{r^2}\left(2r\frac{\partial v}{\partial r} + r^2\frac{\partial^2 v}{\partial r^2}\right) = \frac{1}{c^2}\frac{\partial^2 v}{\partial t^2}.$$

We have the identity

$$\frac{\partial^2(rv)}{\partial r^2} \equiv \frac{\partial}{\partial r}\left(v + r\frac{\partial v}{\partial r}\right) \equiv \frac{\partial v}{\partial r} + \frac{\partial v}{\partial r} + r\frac{\partial^2 v}{\partial r^2} \equiv \frac{1}{r}\left(2r\frac{\partial v}{\partial r} + r^2\frac{\partial^2 v}{\partial r^2}\right).$$

Thus the wave equation takes the form

$$\frac{1}{r}\frac{\partial^2(rv)}{\partial r^2} = \frac{1}{c^2}\frac{\partial^2 v}{\partial t^2}$$

or

$$\frac{\partial^2(rv)}{\partial r^2} = \frac{1}{c^2}\frac{\partial^2(rv)}{\partial t^2}.$$

We define

$$\psi(r,t) := rv(r,t).$$

Then

$$\frac{\partial^2\psi}{\partial r^2} = \frac{1}{c^2}\frac{\partial^2\psi}{\partial t^2}.$$

We have found the general solution of this partial differential equation, namely

$$\psi(r,t) = f_1(r - ct) + f_2(r + ct).$$

It follows that

$$v(r,t) = \frac{1}{r}f_1(r - ct) + \frac{1}{r}f_2(r + ct)$$

with $r > 0$. Let

$$v_0(r) = v(r,0), \qquad v_1(r) = \frac{\partial v(r,0)}{\partial t}$$

be the initial distributions. Then

$$\psi_0(r) = \psi(r,0) = (rv)(r,0) = rv_0$$

$$\psi_1(r) = \frac{\partial\psi}{\partial t}(r,0) = r\frac{\partial v}{\partial t}(r,0) = rv_1(r).$$

Using the method described above the solution to the initial value problem is given by

$$v(r,t) = \frac{1}{2r}(r+ct)v_0(r+ct) + \frac{1}{2r}(r-ct)v_0(r-ct) + \frac{1}{2cr}\int_{r-ct}^{r+ct} \alpha v_1(\alpha)d\alpha.$$

Problem 13. Let

$$-\frac{\hbar}{i}\frac{\partial\psi}{\partial t} = \hat{H}\psi \tag{1}$$

be the *Schrödinger equation*, where

$$\hat{H} := -\frac{\hbar^2}{2m}\Delta + U(\mathbf{r})$$

$$\Delta := \frac{\partial^2}{\partial x_1^2} + \frac{\partial^2}{\partial x_2^2} + \frac{\partial^2}{\partial x_3^2}$$

and $\mathbf{r} = (x_1, x_2, x_3)$. Let

$$\rho(\mathbf{r},t) := \bar{\psi}(\mathbf{r},t)\psi(\mathbf{r},t)$$

where $\bar{\psi}$ denotes the complex conjugate of ψ. Find $\mathbf{j}$ such that

$$\mathrm{div}\mathbf{j} + \frac{\partial\rho}{\partial t} = 0 \tag{2}$$

where

$$\mathrm{div}\mathbf{j} := \frac{\partial j_1}{\partial x_1} + \frac{\partial j_2}{\partial x_2} + \frac{\partial j_3}{\partial x_3}.$$

Equation (2) is called a *conservation law*.

Solution 13. Since

$$-\frac{\hbar}{i}\frac{\partial\psi}{\partial t} = -\frac{\hbar^2}{2m}\Delta\psi + U\psi$$

we obtain

$$\frac{\hbar}{i}\frac{\partial\bar{\psi}}{\partial t} = -\frac{\hbar^2}{2m}\Delta\bar{\psi} + U\bar{\psi}.$$

From ρ we obtain

$$\frac{\partial\rho}{\partial t} = \frac{\partial}{\partial t}(\bar{\psi}\psi) = \bar{\psi}\frac{\partial\psi}{\partial t} + \frac{\partial\bar{\psi}}{\partial t}\psi.$$

Inserting the Schrödinger equation gives

$$-\frac{\hbar}{i}\frac{\partial\rho}{\partial t} = \bar{\psi}\hat{H}\psi - \psi\hat{H}\bar{\psi}.$$

Now

$$\bar{\psi}\hat{H}\psi - \psi\hat{H}\bar{\psi} = -\frac{\hbar^2}{2m}(\bar{\psi}\Delta\psi - \psi\Delta\bar{\psi}) = -\frac{\hbar^2}{2m}\operatorname{div}(\bar{\psi}\nabla\psi - \psi\nabla\bar{\psi}).$$

Here ∇ denotes the gradient. Thus

$$\mathbf{j} = \frac{\hbar}{2mi}(\bar{\psi}\nabla\psi - \psi\nabla\bar{\psi}).$$

Chapter 14

Integration

The fundamental theorem of calculus states that if a function f is Riemann integrable on the interval $I = [a, b]$ and g is a differentiable function on (a, b) such that $dg/dx = f(x)$ for all $x \in (a, b)$ and $\lim_{x \to a^+} g(x)$, $\lim_{x \to b^-} g(x)$ exist, then

$$\int_a^b f(x)dx = \lim_{x \to b^-} g(x) - \lim_{x \to a^+} g(x).$$

If f and g are differentiable functions on the interval $[a, b]$ and if the function df/dx and dg/dx are Riemann integrable on $[a, b]$ then (*integration by parts*)

$$\int_a^b f(x) \frac{dg}{dx} dx = f(b)g(b) - f(a)g(a) - \int_a^b \frac{df}{dx} g(x)dx.$$

Every function which is Riemann integrable is Lebesgue integrable and the values of the two integrals are equal. A standard example of a function on an interval $[a, b] \subset \mathbb{R}$ which is Lebesgue integrable but not Riemann integrable is the function $\chi_{\mathbb{Q}}$, where χ is the indicator function.

Problem 1. Calculate

$$I = \int_0^\infty e^{-x^2} dx.$$

Solution 1. To evaluate the integral we transform the single integral into a double integral. Thus

$$I^2 = \int_0^\infty \left(\int_0^\infty e^{-x^2} dx \right) e^{-y^2} dy$$

$$= \int_0^\infty \int_0^\infty e^{-x^2} e^{-y^2} \, dx dy$$

$$= \int_0^\infty \int_0^\infty e^{-(x^2+y^2)} \, dx dy.$$

Introducing *polar coordinates*

$$x(r, \phi) = r \cos \phi, \qquad y(r, \phi) = r \sin \phi$$

with $0 \leq \phi < \pi/2$, $0 \leq r < \infty$ we obtain

$$I^2 = \int_0^{\pi/2} \int_0^\infty e^{-r^2} r \, dr d\phi = \int_0^{\pi/2} -\frac{1}{2} e^{-r^2} \Big]_0^\infty d\phi = \frac{1}{2} \int_0^{\pi/2} d\phi = \frac{1}{4}\pi.$$

It follows that $I = \sqrt{\pi}/2$.

Problem 2.   Calculate the integral

$$I = \int_0^{\pi/2} \sin^{2n}(\theta) \cos^{2n+1}(\theta) d\theta$$

by considering the double integral

$$\int \int_D (r \sin(\theta))^{2n} (r \cos(\theta))^{2n+1} e^{-r^2} r \, dr d\theta$$

where D is the first quadrant.

Solution 2.   We can write

$$I \int_0^\infty r^{4n+2} e^{-r^2} dr = \int \int_D (r \sin(\theta))^{2n} (r \cos(\theta))^{2n+1} e^{-r^2} r \, dr d\theta$$

$$= \left(\int_0^\infty y^{2n} e^{-y^2} dy \right) \left(\int_0^\infty x^{2n+1} e^{-x^2} dx \right).$$

Changing all integration variables to s, we have

$$I = \frac{\int_0^\infty s^{2n} e^{-s^2} ds \int_0^\infty s^{2n+1} e^{-s^2} ds}{\int_0^\infty s^{4n+2} e^{-s^2} ds} = \frac{I_{2n} I_{2n+1}}{I_{4n+2}}$$

where

$$I_k = \int_0^\infty s^k e^{-s^2} ds.$$

Now we have $I_1 = 1/2$ and integration by parts provides

$$I_k = \frac{k-1}{2} I_{k-2}, \quad k \geq 3.$$

It follows that

$$I = \frac{I_{2n} \cdot n \cdot (n-1) \cdots 1 \cdot I_1}{\frac{4n+1}{2} \cdot \frac{4n-1}{2} \cdots \frac{2n+1}{2} \cdot I_{2n}} = \frac{n!2^n}{(4n+1)(4n-1)\cdots(2n+1)}.$$

Problem 3. Given that

$$\int_0^\infty \frac{\sin x}{x} dx = \frac{1}{2}\pi.$$

Find

$$I = \int_0^\infty \frac{\sin^2 x}{x^2} dx.$$

Solution 3. We calculate the more general integral

$$I(\epsilon) = \int_0^\infty \frac{\sin^2(\epsilon x)}{x^2} dx, \qquad \epsilon \geq 0$$

by using a technique called *parameter differentiation*. Differentiating each side of the previous equation with respect to ϵ (we can interchange differentiation and integration), we obtain

$$\frac{dI(\epsilon)}{d\epsilon} = \int_0^\infty \frac{2x \sin(\epsilon x) \cos(\epsilon x)}{x^2} dx = \int_0^\infty \frac{\sin(2\epsilon x)}{x} dx.$$

We set $y = 2\epsilon x$. Therefore we find $dy = 2\epsilon dx$, and

$$\frac{dI(\epsilon)}{d\epsilon} = \int_0^\infty \frac{\sin y}{y} dy = \frac{1}{2}\pi.$$

Integrating each side gives $I(\epsilon) = \frac{1}{2}\pi\epsilon + C$, where C is the constant of integration. Since $I(0) = 0$, we obtain $C = 0$. Thus

$$I(\epsilon) = \frac{1}{2}\pi\epsilon \qquad \epsilon \geq 0.$$

Setting $\epsilon = 1$ yields $I(1) = I = \pi/2$.

Problem 4. (i) Let $\alpha_j \in \mathbb{C}$. Calculate

$$I = \int_0^\infty d\tau_2 \int_0^{\tau_2} d\tau_1 \exp(\alpha_1\tau_1 + \alpha_2\tau_2) \tag{1}$$

where $\Re\alpha_2 < 0$ and $\Re(\alpha_1 + \alpha_2) < 0$. Here $\Re$ denotes the real part.

(ii) Calculate

$$I = \int_0^\infty d\tau_n \int_0^{\tau_n} d\tau_{n-1} \cdots \int_0^{\tau_2} d\tau_1 \exp\left(\sum_{m=1}^n \alpha_m \tau_m\right) \tag{2}$$

where

$$\Re \sum_{m=k}^n \alpha_m < 0$$

for all $k \le n$.

Solution 4. (i) Equation (1) can be written in the form

$$\int_0^\infty d\tau_2 \int_0^{\tau_2} d\tau_1 \exp(\alpha_1 \tau_1 + \alpha_2 \tau_2) = \int_0^\infty d\tau_2 \int_0^\infty d\tau_1 \Theta(\tau_2 - \tau_1) \exp(\alpha_1 \tau_1 + \alpha_2 \tau_2)$$

where

$$\Theta(\tau_2 - \tau_1) = \begin{cases} 1 & \text{for } \tau_2 > \tau_1 \\ 0 & \text{otherwise.} \end{cases}$$

Θ is called the *step function*. By the substitution $x_1 = \tau_1$, $x_2 = \tau_2 - \tau_1$ we obtain

$$I = \left(\int_0^\infty dx_2 e^{\alpha_2 x_2}\right)\left(\int_0^\infty dx_1 e^{(\alpha_1 + \alpha_2)x_1}\right) = \frac{1}{\alpha_2(\alpha_1 + \alpha_2)}.$$

(ii) By the substitution

$$x_{n-1} = \tau_{n-1}, \qquad x_n = \tau_n - \tau_{n-1}$$

we write (2) in the form

$$I = \left(\int_0^\infty dx_n \exp(\alpha_n x_n)\right)\left(\int_0^\infty dx_{n-1} \int_0^{x_{n-1}} d\tau_{n-2} \cdots \int_0^{\tau_2} d\tau_1\right.$$
$$\left. \times \exp\left(\sum_{m=1}^{n-2} \alpha_m \tau_m\right) \exp((\alpha_n + \alpha_{n-1})x_{n-1})\right).$$

Repeating this process we can write the integral

$$I = \left(\int_0^\infty dx_n \exp(\alpha_n x_n)\right)\left(\int_0^\infty dx_{n-1} \exp(\alpha_n + \alpha_{n-1})x_{n-1}\right) \cdots$$
$$\times \left(\int_0^\infty dx_1 \exp\left(\left(\sum_{m=1}^n \alpha_m\right)x_1\right)\right).$$

We obtain

$$I = (-1)^n (\alpha_n)^{-1} (\alpha_n + \alpha_{n-1})^{-1} \cdots \left(\sum_{m=1}^n \alpha_m\right)^{-1}$$

since for $s < 0$ we have

$$\int_0^\infty e^{sx} dx = \frac{1}{s} e^{sx} \Big|_0^\infty = -\frac{1}{s}.$$

Problem 5.   Let A be an $n \times n$ hermitian matrix. Prove the identity

$$e^{A^2} \equiv \int_{-\infty}^{+\infty} \exp(-\pi x^2 I_n - 2Ax\sqrt{\pi}) dx \tag{1}$$

where I_n is the $n \times n$ unit matrix.

Solution 5.   Since A is an $n \times n$ hermitian matrix there is an $n \times n$ unitary matrix U such that the matrix U^*AU is diagonal, where $U^* = U^{-1}$. We write

$$\tilde{A} = U^*AU = \text{diag}(\lambda_1, \lambda_2, \ldots, \lambda_n). \tag{2}$$

Obviously, $\lambda_1, \ldots, \lambda_n$ are the eigenvalues of the matrix A. Since A is hermitian the eigenvalues are real. From (1) we obtain

$$U^* e^{A^2} U \equiv U^* \left(\int_{-\infty}^{+\infty} dx \exp(-\pi x^2 I - 2Ax\sqrt{\pi}) \right) U. \tag{3}$$

Since

$$U^* e^{A^2} U \equiv e^{U^* A^2 U} \equiv e^{U^* AUU^* AU} \equiv e^{\tilde{A}^2}$$

and

$$U^* \left(\int_{-\infty}^{+\infty} \exp(-\pi x^2 I_n - 2Ax\sqrt{\pi}) \right) U = \int_{-\infty}^{+\infty} \exp(-\pi x^2 I_n - 2\tilde{A}x\sqrt{\pi}) dx$$

we obtain from identity (3) that

$$e^{\tilde{A}^2} \equiv \int_{-\infty}^{+\infty} \exp(-\pi x^2 I - 2\tilde{A}x\sqrt{\pi}) dx.$$

From (2) we obtain

$$\tilde{A}^2 = \text{diag}(\lambda_1^2, \lambda_2^2, \ldots, \lambda_n^2)$$

and

$$e^{\tilde{A}^2} = \text{diag}(e^{\lambda_1^2}, e^{\lambda_2^2}, \ldots, e^{\lambda_n^2}).$$

Thus

$$\int_{-\infty}^{+\infty} \exp(-\pi x^2 I - 2\tilde{A}x\sqrt{\pi}) dx$$

$$= \int_{-\infty}^{+\infty} dx \exp\left(\operatorname{diag}\left(-\pi x^2 - 2\lambda_1 x\sqrt{\pi}, \ldots, -\pi x^2 - 2\lambda_n x\sqrt{\pi}\right)\right)$$

$$= \int_{-\infty}^{+\infty} dx \operatorname{diag}\left(e^{-\pi x^2 - 2\lambda_1 x\sqrt{\pi}}, \ldots, e^{-\pi x^2 - 2\lambda_n x\sqrt{\pi}}\right)$$

$$= \operatorname{diag}\left(\int_{-\infty}^{+\infty} e^{-\pi x^2 - 2\lambda_1 x\sqrt{\pi}} dx, \ldots, \int_{-\infty}^{+\infty} e^{-\pi x^2 - 2\lambda_n x\sqrt{\pi}} dx\right).$$

Since

$$\int_{-\infty}^{+\infty} e^{-\pi x^2 - 2\lambda_i x\sqrt{\pi}} dx = \exp(\lambda_i^2)$$

we have

$$\int_{-\infty}^{+\infty} \exp(-\pi x^2 I_n - 2\tilde{A}x\sqrt{\pi}) dx \equiv \operatorname{diag}(e^{\lambda_1^2}, e^{\lambda_2^2}, \ldots, e^{\lambda_n^2}).$$

This proves the identity (3). Since $U^* = U^{-1}$ we also proved identity (1).

Problem 6.   Given the nonlinear system of differential equations

$$\frac{d^2\theta}{ds^2} - \frac{2b\sin\phi}{a + b\cos\phi}\frac{d\theta}{ds}\frac{d\phi}{ds} = 0, \qquad \frac{d^2\phi}{ds^2} + \frac{(a + b\cos\phi)\sin\phi}{b}\left(\frac{d\theta}{ds}\right)^2 = 0 \quad (1)$$

where $a > b > 0$. Show that

$$\frac{d^2\phi}{ds^2} + \frac{C^2\sin\phi}{b(a + b\cos\phi)^3} = 0 \tag{2}$$

where C is a constant of integration.

Solution 6.   From (1) we find

$$\frac{\frac{d^2\theta}{ds^2}}{\frac{d\theta}{ds}} = \frac{2b\sin\phi}{a + b\cos\phi}\frac{d\phi}{ds}$$

or

$$\frac{d}{ds}\left(\ell n\left(\frac{d\theta}{ds}\right)\right) = \frac{2b\sin\phi}{a + b\cos\phi}\frac{d\phi}{ds}.$$

Therefore

$$\int \frac{d}{ds}\left(\ell n\frac{d\theta}{ds}\right) ds = \int \frac{2b\sin\phi}{a + b\cos\phi}\frac{d\phi}{ds} ds = \int \frac{2b\sin\phi}{a + b\cos\phi} d\phi$$

or

$$\ell n\left(\frac{d\theta}{ds}\right) = -2\ell n|a + b\cos\phi| + K$$

where K is the constant of integration. From this equation we obtain

$$\frac{d\theta}{ds} = \frac{C}{(a + b\cos\phi)^2}$$

where $C = e^K$. Inserting this equation into (1b) yields (2).

Problem 7. Calculate

$$I(\lambda) = \int_{|\mathbf{q}| < k_F} \frac{dq_1 \, dq_2 \, dq_3}{|\mathbf{k} - \mathbf{q}|^2 + \lambda^2} \tag{1}$$

where

$$|\mathbf{k} - \mathbf{q}|^2 = (k_1 - q_1)^2 + (k_2 - q_2)^2 + (k_3 - q_3)^2$$

and $\lambda^2 > 0$.

Solution 7. Introducing spherical coordinates (q, ϕ, θ) we obtain

$$I(\lambda) = \int_0^{k_F} q^2 dq \int_0^\pi \sin\theta d\theta \int_0^{2\pi} d\phi \frac{1}{k^2 - 2qk\cos\theta + q^2 + \lambda^2}$$

where θ is given by

$$\mathbf{k} \cdot \mathbf{q} = kq\cos\theta.$$

We set $u = \cos\theta$. Then $du = -\sin\theta d\theta$ and $\theta = 0 \to u = 1, \theta = \pi \to u = -1$. We obtain

$$I(\lambda) = -2\pi \int_0^{k_F} q^2 dq \int_1^{-1} du \frac{1}{k^2 + q^2 - 2qku + \lambda^2}.$$

The integration with respect to u can easily be performed. We find

$$I(\lambda) = 2\pi \int_0^{k_F} q^2 dq \left[\frac{1}{-2qk} \ln(k^2 + q^2 + \lambda^2 - 2qku) \right]_{u=1}^{u=-1}.$$

Thus

$$I(\lambda) = -\frac{\pi}{k} \int_0^{k_F} q \left(\ln(k^2 + q^2 + \lambda^2 + 2qk) - \ln(k^2 + q^2 + \lambda^2 - 2qk) \right) dq$$

and

$$I(\lambda) = -\frac{\pi}{k} \underbrace{\int_0^{k_F} q \ln(\lambda^2 + (k+q)^2) dq}_{F_1} + \frac{\pi}{k} \underbrace{\int_0^{k_F} q \ln(\lambda^2 + (k-q)^2) dq}_{F_2}.$$

Now we have to calculate the integrals F_1 and F_2. We find

$$F_1 = \int_{q=0}^{k_F} q \ln(\lambda^2 + (k+q)^2) dq = \int_{x=k}^{k_F+k} (x-k) \ln(\lambda^2 + x^2) dx$$

or

$$F_1 = -\underbrace{\int_{x=k}^{k_F+k} k \ln(\lambda^2 + x^2) dx}_{F_1'} + \underbrace{\int_{x=k}^{k_F+k} x \ln(\lambda^2 + x^2) dx}_{F_1''} .$$

Now we have to calculate F_1' and F_1''. Since

$$\int \ln(\lambda^2 + x^2) dx = x \ln(x^2 + \lambda^2) + 2\lambda \arctan \frac{x}{\lambda} - 2x$$

we find for F_1'

$$F_1' = -k[(k_F + k) \ln((k_F + k)^2 + \lambda^2) + 2\lambda \arctan \left(\frac{k_F + k}{\lambda} \right) - 2(k_F + k)$$

$$-k \ln(k^2 + \lambda^2) - 2\lambda \arctan \left(\frac{k}{\lambda} \right) + 2k] .$$

Since

$$\int x f(x^2) dx = \frac{1}{2} \int f(u) du$$

with $x^2 = u$ we find

$$F_1'' = \int_{x=k}^{k_F+k} x \ln(\lambda^2 + x^2) dx = \frac{1}{2} \int_{u=k^2}^{(k_F+k)^2} \ln(\lambda^2 + u) du.$$

Thus

$$F_1'' = \frac{1}{2}(((k_F + k)^2 + \lambda^2) \ln((k_F + k)^2 + \lambda^2) - (k_F + k)^2$$

$$-(k^2 + \lambda^2) \ln(k^2 + \lambda^2) + k^2).$$

The integral F_2 is given by

$$F_2 = \int_{q=0}^{k_F} q \ln(\lambda^2 + (q-k)^2) dq.$$

It follows that

$$F_2 = \int_{x=-k}^{k_F-k} (x+k) \ln(\lambda^2 + x^2) dx$$

$$= \underbrace{\int_{x=-k}^{k_F-k} k \ln(\lambda^2 + x^2) dx}_{F_2'} + \underbrace{\int_{x=-k}^{k_F-k} x \ln(\lambda^2 + x^2) dx}_{F_2''} .$$

Now we have to calculate F_2'. We obtain

$$F_2' = \int_{x=-k}^{k_F-k} k \ln(\lambda^2 + x^2)dx.$$

Thus

$$F_2' = k(k_F - k)\ln((k_F - k)^2 + \lambda^2) + 2k\lambda \arctan\left(\frac{k_F - k}{\lambda}\right)$$

$$-2k(k_F - k) + k^2 \ln(k^2 + \lambda^2) - 2k\lambda \arctan\left(\frac{-k}{\lambda}\right) - 2k^2.$$

For F_2'' we obtain

$$F_2'' = \int_{x=-k}^{k_F-k} x \ln(\lambda^2 + x^2)dx = \frac{1}{2}\int_{u=k^2}^{(k_F-k)^2} \ln(\lambda^2 + u)du$$

or

$$F_2'' = \frac{1}{2}(((k_F - k)^2 + \lambda^2)\ln((k_F - k)^2 + \lambda^2) - (k_F - k)^2 - (k^2 + \lambda^2)\ln(k^2 + \lambda^2) + k^2)$$

where we have used $u = x^2$ and therefore $du = 2xdx$. The integral (1) plays an important role in solid state physics. Here $\hbar k_f$ is the *Fermi momentum*. The absolute value of momentum of particles at zero temperature is called the Fermi momentum and the region of momentum space with momentum $\hbar k_f$ is called the Fermi surface.

Problem 8. Calculate the integral (which plays a rôle in electrostatic fields)

$$U(r) = \frac{1}{4\pi}\int_{r'}\int_{\theta=0}^{\pi}\int_{\phi=0}^{2\pi} \frac{\rho(r')}{(r^2 + r'^2 - 2rr'\cos\theta)^{1/2}} r'^2 \sin\theta dr' d\theta d\phi.$$

Solution 8. The integration with respect to ϕ can easily be performed. We find

$$U(r) = \frac{1}{2}\int_{r'}\int_{\theta=0}^{\pi} \frac{\rho(r')}{(r^2 + r'^2 - 2rr'\cos\theta)^{1/2}} r'^2 \sin\theta dr' d\theta.$$

To perform the θ-integration we set $u = \cos\theta$. Therefore, $du = -\sin\theta d\theta$ and

$$U(r) = -\frac{1}{2}\int_{r'}\int_{u=1}^{-1} \frac{\rho(r')}{(r^2 + r'^2 - 2rr'u)^{1/2}} r'^2 dr' du.$$

Thus

$$U(r) = \frac{1}{2} \int_{r'} \rho(r') \left(\frac{(r^2 + r'^2 + 2rr')^{1/2}}{rr'} - \frac{(r^2 + r'^2 - 2rr')^{1/2}}{rr'} \right) r'^2 dr'.$$

Since

$$(r^2 + r'^2 - 2rr')^{1/2} = ((r - r')^2)^{1/2} = r - r' \quad \text{for} \quad r \geq r'$$

$$(r^2 + r'^2 - 2rr')^{1/2} = ((r' - r)^2)^{1/2} = r' - r \quad \text{for} \quad r' \geq r$$

we obtain

$$U(r) = \frac{1}{2} \int_{r'=0}^{r} \rho(r') \frac{(r + r') - (r - r')}{r} r' dr'$$
$$+ \frac{1}{2} \int_{r'=r}^{\infty} \rho(r') \frac{(r + r') - (r' - r)}{r} r' dr'.$$

Consequently,

$$U(r) = \frac{1}{r} \int_{r'=0}^{r} \rho(r') r'^2 dr' + \int_{r'=r}^{\infty} \rho(r') r' dr'.$$

Problem 9.   Two friends plan to meet at the library during a given 1-hour period. Their arrival times are independent and randomly distributed across the 1-hour period. Each agrees to wait for 15 minutes, or until the end of the hour. If the friend has not appeared during that time, the other friend will leave. Find the probability that the friends will meet.

Solution 9.   If X_1 denotes one person's arrival time in $[0,1]$, the 1-hour period, and X_2 denotes the second person's arrival time, then (X_1, X_2) can be modeled as having a two-dimensional uniform distribution over the unit square. That is,

$$f(x_1, x_2) = \begin{cases} 1 & 0 \leq x_1 \leq 1, \ 0 \leq x_2 \leq 1 \\ 0 & \text{otherwise}. \end{cases} \tag{1}$$

The event that the two friends will meet depends upon the time, U, between their arrivals, where
$$U = |X_1 - X_2|.$$

Next we find the distribution function for the random variable U. For $U \leq u$ we have $|X_1 - X_2| \leq u$. Therefore
$$-u \leq X_1 - X_2 \leq u.$$

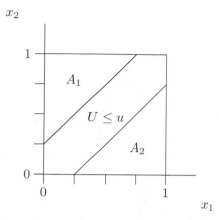

The figure shows that the region over which (X_1, X_2) has positive probability is the region defined by $U \leq u$. The probability that $U \leq u$ can be found by integrating f over the six-sided region shown in the figure. This can be simplified by integrating over the triangles (A_1 and A_2) and subtracting from one. We have

$$F_U(u) = P(U \leq u) = \int\int_{|x_1 - x_2| \leq u} f(x_1, x_2) dx_1 dx_2.$$

Thus

$$F_U(u) = 1 - \int\int_{A_1} f(x_1, x_2) dx_1 dx_2 - \int\int_{A_2} f(x_1, x_2) dx_1 dx_2.$$

Therefore

$$F_U(u) = 1 - \int_u^1 \int_0^{x_2 - u} dx_1 dx_2 - \int_0^{1-u} \int_{x_2+u}^1 dx_1 dx_2.$$

Finally, we arrive at

$$F_U(u) = 1 - (1 - u)^2, \quad 0 \leq u \leq 1.$$

Thus

$$P\left(U \leq \frac{15}{60}\right) = 1 - \left(1 - \frac{15}{60}\right)^2 = \frac{7}{16}.$$

Problem 10.   Let $a, b \in \mathbb{R}$, $a, b \neq 0$ and $b > a$. Show that

$$\frac{1}{ab} = \int_0^1 \frac{dz}{(az + b(1 - z))^2}. \tag{1}$$

If a and b have opposite sign, z is to be considered as a complex variable and the path of the integration must deviate from the real axis so as to avoid the singularity at $z^* = b/(b-a)$.

Solution 10.   Identity (1) is obtained by noting that

$$\frac{1}{ab} = \frac{1}{b-a}\left(\frac{1}{a} - \frac{1}{b}\right) = \frac{1}{b-a}\int_a^b \frac{dx}{x^2}$$

and introducing in the last integral the new variable z via $x = az + b(1-z)$. Thus $dx = (a-b)dz$ and

$$z = 0 \Rightarrow x = b, \qquad z = 1 \Rightarrow x = a.$$

The identity (1) holds for all values of a and b with $a, b \neq 0$. If, however, a and b are of the opposite sign, z is to be considered as a complex variable and the path of integration must deviate from the real axis so as to avoid the singularity (pole) at

$$z^* = \frac{b}{b-a}.$$

Any path joining $z = 0$ and $z = 1$ but not passing through the singularity may actually be chosen since the residue of the integrand at the singularity vanishes.

The integral given by identity (1) is a special case of the following general identity

$$\frac{1}{a_1 a_2 \cdots a_n} = (n-1)! \int_0^1 \cdots \int_0^1 \frac{dz_1 dz_2 \cdots dz_n}{(a_1 z_1 + a_2 z_2 + \cdots + a_n z_n)^n}$$

where

$$\sum_{i=1}^n z_i = 1.$$

To evaluate this integral we can use

$$\frac{1}{(n-1)!} \frac{1}{a_1 a_2 \cdots a_n}$$

$$= \int_0^1 dz_1 \int_0^{z_1} dz_2 \cdots \int_0^{z_{n-2}} dz_{n-1}$$

$$\times \frac{1}{(a_n z_{n-1} + a_{n-1}(z_{n-2} - z_{n-1}) + \cdots + a_1(1 - z_1))^n}$$

$$= \int_0^1 \epsilon_1^{n-2} d\epsilon_1 \int_0^1 \epsilon_2^{n-3} d\epsilon_2 \cdots \int_0^1 d\epsilon_{n-1}$$

$$\times \frac{1}{(a_1 \epsilon_1 \epsilon_2 \cdots \epsilon_{n-1} + a_2 \epsilon_1 \cdots \epsilon_{n-2}(1 - \epsilon_{n-1}) + \cdots + a_n(1 - \epsilon_1))^n}$$

Problem 11. Consider a right circular cone (elliptic cone) of radius r and height h with axis as z-axis and $z \geq 0$. The surface is described by

$$\frac{x^2}{r^2} + \frac{y^2}{r^2} = \frac{z^2}{h^2}. \tag{1}$$

Find the volume of the cone.

Solution 11. To find the volume we have to do the *triple integral*

$$\int_{z=0}^{z=h} \left(\int_{y=g_1(z)}^{y=g_2(z)} \left(\int_{x=f_1(y,z)}^{x=f_2(y,z)} dx \right) dy \right) dz.$$

We consider the quadrant $x \geq 0$ and $y \geq 0$ and we multiply the final result by 4 to find the volume. For the x-integration we have

$$f_1(y,z) = 0, \qquad f_2(y,z) = \frac{\sqrt{r^2 z^2 - h^2 y^2}}{h}$$

where we used (1). Thus

$$\int_{x=0}^{x=\sqrt{r^2 z^2 - h^2 y^2}/h} dx = \frac{1}{h}\sqrt{r^2 z^2 - h^2 y^2}.$$

For the y-integration we have $g_1(z) = 0$ and $g_2(z) = zr/h$ since z and y are related by the linear equation $z = my$. For $z = h$ we have $y = r$ and therefore $m = h/r$, $z = hy/r$. Thus

$$\frac{1}{h} \int_{y=0}^{y=zr/h} \sqrt{r^2 z^2 - h^2 y^2}\, dy$$

To perform the y-integration we apply the substitution $\widetilde{y} = hy$. Therefore $d\widetilde{y} = h\,dy$. For $y = 0$ we have $\widetilde{y} = 0$ and for $y = zr/h$ we obtain $\widetilde{y} = zr$. It follows that

$$\frac{1}{h^2} \int_{\widetilde{y}=0}^{\widetilde{y}=zr} \sqrt{r^2 z^2 - \widetilde{y}^2}\, d\widetilde{y} = \frac{1}{2h^2} \left| \widetilde{y}\sqrt{r^2 z^2 - \widetilde{y}^2} + r^2 z^2 \arcsin\left(\frac{\widetilde{y}}{rz}\right) \right|_{\widetilde{y}=0}^{\widetilde{y}=zr}$$

$$= \frac{1}{4h^2} r^2 z^2 \pi$$

where we used $\arcsin 1 = \pi/2$ and $\arcsin 0 = 0$. The last integration over z is

$$\frac{\pi r^2}{4h^2} \int_{z=0}^{z=h} z^2\, dz = \frac{\pi}{4} \cdot \frac{r^2 h}{3}.$$

Since we only considered the quadrant $x \geq 0$ and $y \geq 0$ we find that the volume of the cone is $V = \pi r^2 h/3$.

Problem 12.   The *Riemann-Liouville definition* for the *fractional deriva-tive* of a function f is given by

$$\frac{d^\alpha f(t)}{dt^\alpha} = \frac{1}{\Gamma(n-\alpha)} \frac{d^n}{dt^n} \int_{\tau=0}^{\tau=t} \frac{f(\tau)}{(t-\tau)^{\alpha-n+1}} d\tau$$

where $\Gamma(.)$ is the gamma function and the integer n is given by $n-1 \le \alpha < n$. Let $f(t) = t^2$. Find the fractional derivative of f with $\alpha = 1/2$.

Solution 12.   Since $\alpha = 1/2$ we have $n = 1$ and $\alpha - n + 1 = 1/2$. Now we have

$$\int_{\tau=0}^{\tau=t} \frac{\tau^2}{\sqrt{t-\tau}} d\tau = -\left. \frac{2(3\tau^2 + 4t\tau + 8t^2)}{15} \sqrt{t-\tau} \right|_{\tau=0}^{\tau=t} = \frac{16}{15} t^{5/2}.$$

Since $dt^{5/2}/dt = 5t^{3/2}/2$ and $\Gamma(1/2) = \sqrt{\pi}$ we obtain

$$\frac{d^{1/2} f(t)}{dt^{1/2}} = \frac{8}{3\sqrt{\pi}} t^{3/2}.$$

Problem 13.   Let

$$x(\tau) = \begin{cases} 1 & \text{for } 0 \le \tau \le 1 \\ 0 & \text{otherwise} \end{cases}$$

and

$$h(\tau) = \begin{cases} 1 & \text{for } 0 \le \tau \le 1 \\ 0 & \text{otherwise} \end{cases}.$$

Find the *convolution integral*

$$y(t) = \int_{-\infty}^{\infty} x(\tau) h(t-\tau) d\tau.$$

Solution 13.   The value of the function y at time t is given by the amount of overlap (i.e. the integral of the overlapping region) between $h(t-\tau)$ and $x(\tau)$. Thus we have four limits of integration, namely $t < 0$, $0 \le t < 1$, $1 \le t \le 2$ and $t > 2$. For the region $t < 0$ we have $y(t) = 0$. For the region $0 \le t < 1$ we have

$$y(t) = \int_0^t d\tau = t.$$

The third region $1 \le t \le 2$ provides

$$y(t) = \int_{t-1}^1 d\tau = 2 - t.$$

The fourth region $t > 2$ provides $y(t) = 0$. Thus

$$y(t) = \begin{cases} 0 & \text{if} \quad t < 0 \\ t & \text{if} \quad 0 \leq t < 1 \\ 2 - t & \text{if} \quad 1 \leq t \leq 2 \\ 0 & \text{if} \quad t > 2 \end{cases}$$

Problem 14.   The sum

$$\lim_{n \to \infty} \frac{1}{n} \sum_{k=1}^{n} \exp\left(2 \cos\left(\frac{k\pi}{n+1}\right)\right)$$

can be cast into the integral

$$\lim_{n \to \infty} \frac{1}{n} \int_{0}^{n} \exp(2 \cos(\pi x)) dx. \tag{1}$$

Calculate this integral.

Solution 14.   We have

$$\lim_{n \to \infty} \frac{1}{n} \int_{0}^{n} \exp(2 \cos(\pi x)) dx = \lim_{n \to \infty} \frac{1}{n} n \int_{0}^{1} \exp(2 \cos(\pi x)) dx$$

$$= \int_{0}^{1} \exp(2 \cos(\pi x)) dx.$$

Using $y = \pi x$ we arrive at

$$\lim_{n \to \infty} \frac{1}{n} \int_{0}^{n} \exp(2 \cos(\pi x)) dx = \frac{1}{\pi} \int_{0}^{\pi} \exp(2 \cos y) dy = \sqrt{\pi} \Gamma(1/2) I_0(2)$$

where Γ is the gamma function and I_0 is the cylinder function.

Problem 15.   Calculate

$$I(x) = PV \int_{-\infty}^{\infty} \frac{e^{-t^2}}{t - x} dt, \qquad x \in \mathbb{R} \tag{1}$$

where the integral is taken in the sense of the *Cauchy principal value*. This is the *Hilbert transform* of $\exp(-t^2)$.

Solution 15.   By definition of the Cauchy principal value integral, we have

$$I(x) = \lim_{\epsilon \to 0} \left(\int_{-\infty}^{x-\epsilon} + \int_{x+\epsilon}^{\infty} \right) \frac{e^{-t^2}}{t - x} dt$$

$$= \lim_{\epsilon \to 0} \int_{\epsilon}^{\infty} \frac{e^{-(x+t)^2} - e^{-(x-t)^2}}{t} \, dt$$

$$= -e^{-x^2} \int_{0}^{\infty} \frac{e^{2xt} - e^{-2xt}}{t} e^{-t^2} \, dt$$

or

$$I(x) = -4xe^{-x^2} \int_{0}^{\infty} \frac{\sinh(2xt)}{2xt} e^{-t^2} \, dt \, . \tag{2}$$

Using Taylor expansion of the hyperbolic sine, term-by-term integration and

$$\int_{0}^{\infty} t^{2k} e^{-t^2} \, dt = \frac{1}{2} \Gamma\left(k + \frac{1}{2}\right)$$

yields

$$\int_{0}^{\infty} \frac{\sinh(2xt)}{2xt} e^{-t^2} \, dt = \frac{1}{2} \sum_{k=0}^{\infty} \frac{\Gamma(k + \frac{1}{2})}{\Gamma(2k + 2)} (2x)^{2k} \, . \tag{3}$$

The duplication formula for the Gamma function is

$$\Gamma(2k + 2) \equiv \frac{2^{2k+3/2}}{\sqrt{2\pi}} \Gamma(k + 1) \Gamma\left(k + \frac{1}{2}\right) \, .$$

Thus

$$\int_{0}^{\infty} \frac{\sinh(2xt)}{2xt} e^{-t^2} \, dt = \frac{\sqrt{\pi}}{4} \sum_{k=0}^{\infty} \frac{1}{k + \frac{1}{2}} \frac{x^{2k}}{k!} \, . \tag{4}$$

The series on the right-hand side is the *error function* of a purely imaginary argument, namely

$$\sum_{k=0}^{\infty} \frac{1}{k + \frac{1}{2}} \frac{x^{2k}}{k!} = \sqrt{\pi} \frac{\mathrm{erf}(ix)}{ix} = \frac{2}{x} \int_{0}^{x} e^{t^2} \, dt \, . \tag{5}$$

Substitution of (4) and (5) into (2) expresses the Hilbert transform in terms of *Dawson's integral*

$$I(x) = -2\sqrt{\pi} F(x), \qquad F(x) := e^{-x^2} \int_{0}^{x} e^{t^2} \, dt \, .$$

From the asymptotic expansion of the error function one finds

$$I(x) \sim -\frac{\sqrt{\pi}}{x} \left(1 + \sum_{k=1}^{\infty} \frac{1 \cdot 3 \cdots (2k-1)}{(2x^2)^k}\right) \qquad \text{as} \qquad x \to \infty \, .$$

Problem 16.   Let $n = 1, 2, \ldots$ and

$$I_n := \int_{0}^{\pi/4} \tan^n(x) \, dx \, .$$

Calculate $f(n) = I_n + I_{n+2}$.

Solution 16. Using $1 + \tan^2(x) \equiv \sec^2(x)$, $\sec(x) \equiv 1/\cos(x)$ and the substitution $u = \tan(x)$ we have

$$
\begin{aligned}
I_n + I_{n+2} &= \int_0^{\pi/4} (\tan^n(x) + \tan^{n+2}(x))dx \\
&= \int_0^{\pi/4} \tan^n(x)\sec^2(x)dx \\
&= \int_0^1 u^n du \\
&= \frac{1}{n+1}.
\end{aligned}
$$

Problem 17. Let f be a Riemann integrable periodic function over the interval $[0, 2\pi]$. Then f can be written as

$$
f(\theta) = \sum_{n=-\infty}^{\infty} c_n e^{in\theta}
$$

with

$$
c_m = \frac{1}{2\pi} \int_0^{2\pi} e^{-im\theta} f(\theta)d\theta.
$$

Let $m \geq 0$ and

$$
s_m := \sum_{n=-m}^{m} c_n e^{in\theta}.
$$

Calculate the *Cesáro sum* $\sigma_m(\theta)$ defined by

$$
\sigma_m(\theta) := \frac{s_0 + s_1 + \cdots + s_m}{m+1}.
$$

Express the result using the *Fejér kernels*

$$
K_m(y) := \sum_{n=-m}^{m} \frac{m+1-|n|}{m+1} e^{iny}.
$$

(ii) Show that

$$
K_m(y) = \frac{1}{m+1} \left(\frac{\sin((m+1)y/2)}{\sin(y/2)} \right)^2.
$$

Solution 17.   (i) We have

$$\sigma_m(\theta) = \frac{1}{m+1} \sum_{j=0}^{m} \sum_{n=-j}^{j} c_n e^{in\theta}$$

$$= \sum_{n=-m}^{m} \frac{m+1-|n|}{m+1} c_n e^{in\theta}$$

$$= \sum_{n=-m}^{m} \frac{m+1-|n|}{m+1} \left(\frac{1}{2\pi} \int_{0}^{2\pi} e^{-inx} f(x) dx \right) e^{in\theta}$$

$$= \frac{1}{2\pi} \int_{0}^{2\pi} f(x) \left(\sum_{n=-m}^{m} \frac{m+1-|n|}{m+1} e^{in(\theta-x)} \right) dx$$

$$= \frac{1}{2\pi} \int_{0}^{2\pi} f(x) K_m(\theta - x) dx \,.$$

The substitution $y = \theta - x$ provides

$$\frac{1}{2\pi} \int_{0}^{2\pi} f(x) K_m(\theta - x) dx = \frac{1}{2\pi} \int_{0}^{2\pi} f(\theta - y) K_m(y) dy \,.$$

(ii) We have

$$K_m(y) = \frac{1}{m+1} (e^{-imy} + 2e^{-i(m-1)y} + \cdots + (m+1)e^0 + \cdots + e^{imy})$$

$$= \frac{1}{m+1} (e^{-imy/2} + e^{-i(m/2-1)y} + \cdots + e^{imy/2})^2$$

$$= \frac{1}{m+1} \left(\frac{e^{i(m+1)y/2} - e^{-i(m+1)y/2}}{e^{iy/2} - e^{-iy/2}} \right)^2$$

$$= \frac{1}{m+1} \left(\frac{\sin((m+1)y/2)}{\sin(y/2)} \right)^2 \,.$$

Thus $K_m(0) = m+1$. The function K_m satisfies $K_m(y) \geq 0$ for all $y \in \mathbb{R}$.

Problem 18.   Let f be a continuous function.
(i) Show that the double integral

$$\int_{0}^{x} d\xi \int_{0}^{\xi} f(t) dt$$

can be expressed by a single integral.
(ii) Show that $(n \geq 2)$

$$\int_{0}^{x} d\xi_1 \int_{0}^{\xi_1} d\xi_2 \cdots \int_{0}^{\xi_{n-1}} f(\xi_n) d\xi_n$$

can be expressed by a single integral.

Solution 18.   (i) We have

$$\int_0^x d\xi \int_0^\xi f(t)dt = \int_0^x dt \int_t^x f(t)d\xi = \int_0^x f(t)dt \int_t^x d\xi = \int_0^x (x-t)f(t)dt \,.$$

(ii) Using the result from (i) and induction we have

$$\int_0^x d\xi_1 \int_0^{\xi_1} d\xi_2 \cdots \int_0^{\xi_{n-1}} f(\xi_n)d\xi_n = \int_0^x \frac{(x-t)^{n-1}}{(n-1)!} f(t)dt \,.$$

For $n = 2$ we have the result of (i).

Problem 19.   Let $T > 0$ and $\omega = 2\pi/T$. Let $m, n \in \mathbb{N}$ and $\alpha, \beta \in \mathbb{R}$. Calculate

$$I(\alpha, \beta) = \frac{1}{T} \int_0^T c_m \sin(m\omega t + \phi_m - \alpha)c_n \sin(n\omega t + \phi_n - \beta)dt$$

Solution 19.   Since we have the identity

$$\sin(m\omega t + \phi_m - \alpha) \sin(n\omega t + \phi_n - \beta) \equiv$$

$$\frac{1}{2}\cos((m-n)\omega t + \phi_m - \phi_n - \alpha + \beta) - \frac{1}{2}\cos((m+n)\omega t + \phi_m + \phi_n - \alpha - \beta)$$

we obtain

$$I(\alpha, \beta) = \begin{cases} \dfrac{1}{2}c_m^2 \cos(\alpha - \beta) & \text{for} \quad m = n \\ \\ 0 & \text{for} \quad m \neq n \end{cases}$$

Problem 20.   Find the integral

$$I(\ell) = \int_0^{2\pi} \left(\int_0^{|\cos(\theta)|(\ell/2)} dp \right) d\theta \,.$$

Solution 20.   We obtain

$$I(\ell) = \int_0^{2\pi} \frac{\ell}{2}|\cos(\theta)|d\theta == \frac{\ell}{2} \int_0^{2\pi} |\cos(\theta)|d\theta = 2\ell \,.$$

Problem 21.   Evaluate the *Lebesgue integral*

$$\int_0^1 x^2 dx. \tag{1}$$

Solution 21.   Let E be the set $E = [0, 1)$. Then E is an elementary set having Lebesgue measure 1. The function $f : \mathbb{R} \to \mathbb{R}$ defined by

$$f(x) = \begin{cases} x^2 & \text{if } x \in E \\ 0 & \text{if } x \notin E \end{cases}$$

is a measurable function since

$$\{ x : f(x) \geq a \}$$

is measurable for all $a \in \mathbb{R}$. We introduce a monotonic increasing sequence of simple functions tending to f as follows. Let

$$Q_{p,s} := \left\{ x : \frac{p-1}{2^s} \leq x < \frac{p}{2^s} \right\}$$

where $p = 1, 2, 4, \ldots, 2^s$, $s = 1, 2, \ldots$. We define

$$f_s(x) := \begin{cases} \left(\dfrac{p-1}{2^s} \right)^2 & \text{for } x \in Q_{p,s} \\ 0 & \text{for } x \in \mathbb{R} \setminus E \end{cases}$$

Then for all $x \in \mathbb{R}$ we have $0 \leq f_s(x) \leq f(x)$. Moreover, for all $x \in \mathbb{R}$, we find $f_{s+1}(x) \geq f_s(x)$. This means f_s is a monotonic increasing function with increasing s. Furthermore

$$0 \leq f(x) - f_s(x) \leq \left(\frac{2^s}{2^s} \right)^2 - \left(\frac{2^s - 1}{2^s} \right)^2 = \frac{2^{2s} - 1}{2^s 2^s} < \frac{2}{2^s}$$

so that

$$f_s(x) \to f(x) \quad \text{as} \quad s \to \infty.$$

Therefore

$$\int_E f_s(x) dx = \frac{1}{2^s} \left(0 + \left(\frac{1}{2^s} \right)^2 + \left(\frac{2}{2^s} \right)^2 + \cdots + \left(\frac{2^s - 1}{2^s} \right)^2 \right).$$

Thus

$$\int_E f_s(x) dx = \frac{1}{n^3} (1 + 2^2 + 3^2 + \cdots + (n-1)^2) = \frac{1}{n^3} \left(\frac{(n-1)n(2n-1)}{6} \right)$$

where we used $n = 2^s$. Therefore

$$\int_E f_s(x) dx \to \frac{1}{3} \quad \text{as} \quad s \to \infty.$$

It follows that

$$\int_0^1 x^2 dx = \frac{1}{3}.$$

Chapter 15

Continuous Fourier Transform

Let $L_2(\mathbb{R})$ be the Hilbert space of the square integrable function over $\mathbb{R}$. Let $f \in L_2(\mathbb{R})$ and $f \in L_1(\mathbb{R})$. The *Fourier transform* $\hat{f}(k)$ of $f(x)$ is defined as

$$\hat{f}(k) := \int_{\mathbb{R}} f(x)e^{ikx}dx.$$

The inverse Fourier transform is given by

$$f(x) = \frac{1}{2\pi} \int_{\mathbb{R}} \hat{f}(k)e^{-ikx}dk.$$

For $L_2(\mathbb{R}^n)$ we have

$$\hat{f}(\mathbf{k}) := \int_{\mathbb{R}^n} f(\mathbf{x})e^{i\mathbf{k}\cdot\mathbf{x}}d\mathbf{x}$$

and the inverse

$$f(\mathbf{x}) = \frac{1}{(2\pi)^n} \int_{\mathbb{R}^n} \hat{f}(\mathbf{k})e^{-i\mathbf{k}\cdot\mathbf{x}}d\mathbf{k}$$

where $d\mathbf{x} = dx_1 \cdots dx_n$, $d\mathbf{k} = dk_1 \cdots dk_n$ and

$$\mathbf{k} \cdot \mathbf{x} := k_1 x_1 + \cdots + k_n x_n$$

is the scalar product. In some one-dimensional applications x would be the time t and k would then be the frequency ω. In applications in three dimensions $\mathbf{k}$ would be the *wave vector* and $\mathbf{x}$ would be the space coordinates. The Fourier transform and its inverse are linear transformations.

Problem 1.   Let $a > 0$. We define

$$f_a(x) = \begin{cases} \dfrac{1}{2a} & \text{for} \quad |x| \leq a \\ 0 & \text{for} \quad |x| > a \end{cases}. \tag{1}$$

(i) Calculate

$$\int_{\mathbb{R}} f_a(x)dx. \tag{2}$$

(ii) Calculate the *Fourier transform* of f_a. Discuss the cases: a large and a small.

Solution 1.   (i) We find

$$\int_{\mathbb{R}} f_a(x)dx = \frac{1}{2a} \int_{-a}^{a} dx = 1.$$

Thus the integral is independent of a.
(ii) We have

$$\hat{f}_a(k) = \int_{-a}^{a} \frac{e^{ikx}}{2a}dx = \frac{1}{2a}\int_{-a}^{a} e^{ikx}dx.$$

Thus

$$\hat{f}_a(k) = \frac{1}{2aik}\left. |e^{ikx}|\right._{-a}^{a}.$$

Finally

$$\hat{f}_a(k) = \frac{1}{ak}\frac{e^{ika} - e^{-ika}}{2i} = \frac{\sin(ak)}{ak}.$$

For $k = 0$ we find (L'Hospital)

$$\hat{f}_a(0) = 1.$$

Problem 2.   Let $N \in \mathbb{N}$ and let

$$V(t) = \begin{cases} V_0 e^{i\omega_0 t} & \text{if} \quad nT \leq t \leq (nT + \tau) \text{ for } n = 0, 1, 2, \ldots, N-1 \\ 0 & \text{otherwise} \end{cases}$$

where V_0 and τ are positive constants. Calculate the Fourier transform.

Solution 2.   For general N the Fourier transform is given by

$$S(\omega) = V_0 \sum_{n=0}^{N-1} \int_{nT}^{nT+\tau} e^{-i\omega_0 t} e^{i\omega t} dt.$$

Thus

$$S(\omega) = V_0 \frac{e^{i(\omega-\omega_0)\tau}}{i(\omega-\omega_0)} \sum_{n=0}^{N-1} e^{in(\omega-\omega_0)T}.$$

The sum over n is a geometric series, which can be evaluated as follows

$$\sum_{n=0}^{N-1} e^{in\alpha} = \frac{1-e^{iN\alpha}}{1-e^{i\alpha}} = e^{i(N-1)\alpha/2} \frac{\sin(N\alpha/2)}{\sin(\alpha/2)}.$$

With $\alpha := (\omega-\omega_0)T$ we obtain

$$S(\omega) = V_0 \frac{e^{i(\omega-\omega_0)\tau}}{i(\omega-\omega_0)} e^{i(N-1)(\omega-\omega_0)/2} \frac{\sin(N(\omega-\omega_0)T/2)}{\sin((\omega-\omega_0)T/2)}.$$

Problem 3. Calculate the Fourier transform of

$$f(\mathbf{r}) = \frac{1}{r^2+\lambda^2}$$

where $\mathbf{r} = (x_1, x_2, x_3)$,

$$\mathbf{r}^2 \equiv r^2 = x_1^2 + x_2^2 + x_3^2$$

and $\lambda^2 > 0$.

Solution 3. The Fourier transform is defined as

$$\hat{f}(\mathbf{k}) := \int_{-\infty}^{+\infty} \int_{-\infty}^{+\infty} \int_{-\infty}^{+\infty} \frac{e^{i\mathbf{k}\cdot\mathbf{r}}}{r^2+\lambda^2} dx_1 dx_2 dx_3$$

where $\mathbf{k}\cdot\mathbf{r} = k_1 x_1 + k_2 x_2 + k_3 x_3$. Introducing spherical coordinates

$$x_1(r,\phi,\theta) = r\cos\phi\sin\theta, \quad x_2(r,\phi,\theta) = r\sin\phi\sin\theta, \quad x_3(r,\phi,\theta) = r\cos\theta$$

we obtain the volume element

$$dx_1 dx_2 dx_3 = r^2 \sin\theta dr d\theta d\phi$$

where $0 \le \phi < 2\pi$, $0 \le \theta < \pi$, and $r \ge 0$. We can set

$$\mathbf{k}\cdot\mathbf{r} = kr\cos\theta$$

where $k = \|\mathbf{k}\| = \sqrt{k_1^2 + k_2^2 + k_3^2}$. Now (1) takes the form

$$\hat{f}(\mathbf{k}) = \int_{r=0}^{\infty} \int_{\theta=0}^{\pi} \int_{\phi=0}^{2\pi} \frac{e^{ikr\cos\theta}}{r^2+\lambda^2} r^2 \sin\theta dr d\theta d\phi.$$

The integration over ϕ can be easily performed and we find

$$\hat{f}(\mathbf{k}) = 2\pi \int_{r=0}^{\infty} \int_{\theta=0}^{\pi} \frac{e^{ikr\cos\theta}}{r^2 + \lambda^2} r^2 \sin\theta dr d\theta.$$

We set $u = \cos\theta$ and therefore $du = -\sin\theta d\theta$. It follows that $\theta = 0 \rightarrow u = 1$ and $\theta = \pi \rightarrow u = -1$. Then $\hat{f}(\mathbf{k})$ takes the form

$$\hat{f}(\mathbf{k}) = -2\pi \int_{r=0}^{\infty} \frac{r^2}{r^2 + \lambda^2} \left(\int_{u=1}^{u=-1} e^{ikru} du \right) dr.$$

The integration over u can easily be performed. We find

$$\int_{1}^{-1} e^{ikru} du = \frac{-2\sin(kr)}{kr}.$$

Thus we arrive at

$$\hat{f}(\mathbf{k}) = \frac{4\pi}{k} \int_{r=0}^{\infty} \frac{r\sin(kr)}{r^2 + \lambda^2} dr.$$

The integral on the right-hand side can be solved by applying the integration in the complex plane. We obtain

$$\hat{f}(\mathbf{k}) = \frac{2\pi^2}{k} e^{-\lambda k}.$$

Problem 4. Let $f, g \in L_2(\mathbb{R})$ and $f, g \in L_1(\mathbb{R})$. Let

$$\hat{f}(k) = \int_{\mathbb{R}} f(x) e^{ikx} dx, \qquad \hat{g}(k) = \int_{\mathbb{R}} g(x) e^{ikx} dx.$$

Show that

$$(f, g) = \frac{1}{2\pi} (\hat{f}, \hat{g}) \tag{1}$$

where $(\,,\,)$ denotes the scalar product in the Hilbert space $L_2(\mathbb{R})$.

Solution 4. We have

$$(f, g) := \int_{\mathbb{R}} \bar{f}(x) g(x) dx = \frac{1}{2\pi} \int_{\mathbb{R}} \bar{f}(x) \left(\int_{\mathbb{R}} \hat{g}(k) e^{-ikx} dk \right) dx.$$

Thus

$$(f, g) = \frac{1}{2\pi} \int_{\mathbb{R}} \hat{g}(k) \left(\int_{\mathbb{R}} \bar{f}(x) e^{-ikx} dx \right) dk = \frac{1}{2\pi} \int_{\mathbb{R}} \overline{\hat{f}}(k) \hat{g}(k) dk.$$

Thus (1) follows. Equation (1) is called *Parseval's equation*. We have used that

$$\overline{\hat{f}}(k) = \int_{\mathbb{R}} \bar{f}(x) e^{-ikx} dx.$$

Problem 5. The mother *Haar wavelet* is given by

$$f(t) = \begin{cases} -1 & \text{for} \quad 0 \le t < 1/2 \\ +1 & \text{for} \quad 1/2 \le t < 1 \\ 0 & \text{otherwise} \end{cases}$$

Find the Fourier transform

$$\hat{f}(\omega) = \int_{-\infty}^{\infty} f(t)e^{i\omega t}\,dt\,.$$

Solution 5. We obtain

$$\hat{f}(\omega) = 2\frac{1 - \cos(\omega)}{\omega}e^{i(\omega+\pi)/2}$$

with $\hat{f}(0) = 0$ (L'Hospital rule).

Problem 6. The *Poisson wavelet* is given by

$$f(t) = \left(t\frac{d}{dt} + 1 \right) P(t)$$

where

$$P(t) = \frac{1}{\pi}\frac{1}{1 + t^2}\,.$$

Find the Fourier transform of f.

Solution 6. We obtain

$$\hat{f}(\omega) = |\omega|e^{-|\omega|}\,.$$

Thus $\hat{f}$ is not differentiable at $\omega = 0$.

Problem 7. Given the *Mexican hat wavelet*

$$\psi(t) = \frac{2}{\sqrt{3}\sqrt[4]{\pi}}(1 - t^2)e^{-t^2/2} = \frac{2}{\sqrt{3}\sqrt[4]{\pi}}\frac{d}{dt}\left(te^{-t^2/2} \right)\,.$$

Calculate the wavelet transform for the signal

$$f(t) = \begin{cases} 1 & |t| \le \dfrac{1}{2} \\ 0 & \text{otherwise} \end{cases}$$

The *wavelet transform* is given by

$$\mathcal{W}(f(t))(a, b) = \frac{1}{\sqrt{|a|}} \int_{-\infty}^{\infty} f(t)\psi\left(\frac{t - b}{a} \right) dt\,.$$

Solution 7. Using the substitution $\tau = (t-b)/a$ and therefore $d\tau = dt/a$ we find

$$
\mathcal{W}(f(t))(a,b) = \frac{1}{\sqrt{|a|}} \int_{-\infty}^{\infty} f(t)\psi\left(\frac{t-b}{a}\right) dt
$$

$$
= \frac{1}{\sqrt{|a|}} \int_{-1/2}^{1/2} \psi\left(\frac{t-b}{a}\right) dt
$$

$$
= \frac{a}{\sqrt{|a|}} \int_{(-1-2b)/(2a)}^{(1-2b)/(2a)} \psi(\tau)d\tau
$$

$$
= \frac{a}{\sqrt{|a|}} \frac{2}{\sqrt{3}\sqrt[4]{\pi}} \tau e^{-\tau^2/2} \Big|_{(-1-2b)/(2a)}^{(1-2b)/(2a)}
$$

$$
= \frac{1}{\sqrt{3}\sqrt[4]{\pi}} \frac{1}{\sqrt{|a|}} \left((1-2b)e^{-(1-2b)^2/(8a^2)} + (1+2b)e^{-(1+2b)^2/(8a^2)} \right).
$$

Chapter 16

Complex Analysis

Both *Cauchy's integral formula* and the *residue theorem* play a central role in solving problems in complex analysis.

Cauchy's Integral Formula. Let Γ be a simple closed positively oriented contour. If f is analytic in some simply connected domain D containing Γ and z_0 is any point inside Γ, then

$$f(z_0) = \frac{1}{2\pi i} \int_\Gamma \frac{f(z)}{z - z_0} dz \,.$$

Residue Theorem. If Γ is a simple closed positively oriented contour and f is analytic inside and on Γ, except at the points $z_1, z_2, \ldots, z_n$ inside Γ, then

$$\int_\Gamma f(z) dz = 2\pi i \sum_{j=1}^{n} \mathrm{Res}(z_j) \,.$$

If f has a pole of order m at z_0, then

$$\mathrm{Res}(f; z_0) = \lim_{z \to z_0} \frac{1}{(m-1)!} \frac{d^{m-1}}{dz^{m-1}} ((z - z_0)^m f(z)) \,.$$

Suppose that the function f is holomorphic on an open annulus $A(a; r, R)$. Then f has a *Laurent series expansion*

$$f(z) = \sum_{k=-\infty}^{\infty} c_k (z - a)^k$$

which converges absolutely in the annulus and uniformely on compact sub-annuli. The coefficients c_k of the Laurent expansion are determined by uniquely as

$$c_k = \frac{1}{2\pi i} \int_\gamma \frac{f(\zeta)}{\zeta - a} d\zeta, \qquad k = 0, \pm 1, \pm 2, \pm 3, \ldots$$

where γ is any positively oriented Jordan curve in the annulus which wraps around the point a.

Problem 1.   Let $z = x + iy$, $\bar{z} = x - iy$ where $x, y \in \mathbb{R}$.
(i) Show that

$$\frac{\partial}{\partial z} = \frac{1}{2}\left(\frac{\partial}{\partial x} - i\frac{\partial}{\partial y}\right) \tag{1a}$$

$$\frac{\partial}{\partial \bar{z}} = \frac{1}{2}\left(\frac{\partial}{\partial x} + i\frac{\partial}{\partial y}\right). \tag{1b}$$

(ii) Find $dz \wedge d\bar{z}$, where $\wedge$ denotes the exterior product. The exterior product satifies

$$dx \wedge dx = 0, \quad dy \wedge dy = 0, \quad dx \wedge dy = -dy \wedge dx.$$

Solution 1.   (i) Let f be a smooth function of x and y $(x, y \in \mathbb{R})$. Then

$$df = \frac{\partial f}{\partial x}dx + \frac{\partial f}{\partial y}dy. \tag{2}$$

Since

$$dz = d(x + iy) = dx + idy, \qquad d\bar{z} = d(x - iy) = dx - idy$$

we obtain

$$dx = \frac{1}{2}(dz + d\bar{z}), \qquad dy = \frac{1}{2i}(dz - d\bar{z}). \tag{3}$$

Inserting (3) into (2) yields

$$df = \frac{\partial f}{\partial x}\frac{1}{2}(dz + d\bar{z}) + \frac{\partial f}{\partial y}\frac{1}{2i}(dz - d\bar{z})$$

$$= \frac{1}{2}\left(\frac{\partial f}{\partial x} - i\frac{\partial f}{\partial y}\right)dz + \frac{1}{2}\left(\frac{\partial f}{\partial x} + i\frac{\partial f}{\partial y}\right)d\bar{z}.$$

Since

$$df = \frac{\partial f}{\partial z}dz + \frac{\partial f}{\partial \bar{z}}d\bar{z}$$

we find (1a) and (1b). Note that

$$\frac{\partial}{\partial x} = \frac{\partial}{\partial z} + \frac{\partial}{\partial \bar{z}}, \qquad \frac{\partial}{\partial y} = i\left(\frac{\partial}{\partial z} - \frac{\partial}{\partial \bar{z}}\right).$$

(ii) We obtain

$$dz \wedge d\bar{z} = d(x + iy) \wedge d(x - iy) = -2idx \wedge dy.$$

Problem 2.   Discuss the mapping of the periodic strip

$$-\pi < \Re(z) \le +\pi$$

by the function

$$w(z) = \sin(z).$$

Find the images of the straight lines $\Re(z) = $ const and of the straight segments $\Im(z) = $ const.

Solution 2.   Since $z = x + iy$ and

$$w(z(x, y)) = u(x, y) + iv(x, y)$$

where u and v are real-valued functions and

$$\sin(z) = \sin(x + iy) = \frac{e^y + e^{-y}}{2}\sin x + i\frac{e^y - e^{-y}}{2}\cos x$$

we obtain

$$u(x, y) = \frac{e^y + e^{-y}}{2}\sin x, \qquad v(x, y) = \frac{e^y - e^{-y}}{2}\cos x.$$

Let y_0 be a fixed positive number. If $y = y_0$ and x increases from $-\pi$ to π, then the function w describes an ellipse with foci at $\pm i$, since

$$\frac{u^2}{C^2} = \sin^2 x, \qquad \frac{v^2}{D^2} = \cos^2 x$$

and $\sin^2 x + \cos^2 x = 1$ with

$$C = \frac{e^y + e^{-y}}{2}, \qquad D = \frac{e^y - e^{-y}}{2}.$$

The ellipse is described once in the clockwise direction starting at its lowest point. For $y = -y_0$ we obtain the same ellipse described in the opposite direction starting at its highest point. Through each point of the plane (except those along the real segment $[-1, 1]$) passes exactly one such ellipse. If $y_0 = 0$ then

$$u(x, y_0 = 0) = \sin x, \qquad v(x, y_0 = 0) = 0.$$

Consequently, we describe the segment $[-1, 1]$ twice (from 0 to -1, to 0, to 1, to 0). To obtain the complete image of the periodic strip, we take two copies of the w plane, cut one along the positive imaginary axis, the other along the negative imaginary axis and both along the real segment $[-1, 1]$. If we join the banks of the horizontal slits crosswise, we find the two-sheeted Riemann surface which is the one-to-one image of the periodic strips. The lines

$$\Re(z) = \text{const}$$

are taken into hyperbolas orthogonal to the ellipses.

Problem 3.   Calculate

$$\int_0^\infty \frac{x \sin(ax)}{x^2 + \lambda^2} dx$$

where $a, \lambda^2 \in \mathbb{R}$ and $\lambda^2 > 0$.

Solution 3.   We solve the problem by considering the integration in the complex plane. First we notice that

$$\int_0^\infty \frac{x \sin(ax)}{x^2 + \lambda^2} dx = \frac{1}{2} \int_{-\infty}^{+\infty} \frac{x \sin(ax)}{x^2 + \lambda^2} dx = \frac{1}{2} \Im \int_{-\infty}^{+\infty} \frac{x e^{iax}}{x^2 + \lambda^2} dx$$

where $\Im$ denotes the imaginary part. Now we consider

$$\int_C \frac{z e^{iaz}}{z^2 + \lambda^2} dz$$

where C is the contour consisting of the line along the x-axis from $-R$ to $+R$ and the semicircle Γ above the x-axis having this line as diameter. Thus we have

$$\int_C \frac{z e^{iaz}}{z^2 + \lambda^2} dz = \int_{-R}^{+R} \frac{x e^{iax}}{x^2 + \lambda^2} dx + \int_\Gamma \frac{R e^{i\phi} e^{iaR e^{i\phi}}}{R^2 e^{2i\phi} + \lambda^2} R e^{i\phi} i \, d\phi \qquad (1)$$

since $z = R e^{i\phi}$, $dz = R e^{i\phi} i d\phi$. Now

$$z^2 + \lambda^2 \equiv (z + i\lambda)(z - i\lambda)$$

where only the pole at $z = i\lambda$ lies inside C. The left-hand side is calculated with the help of the *residue theorem*. The residue a_1 of a complex function f at $z = a$, where $z = a$ is a pole of order k, is calculated as follows

$$a_{-1} := \lim_{z \to a} \frac{1}{(k-1)!} \frac{d^{k-1}}{dz^{k-1}} \left((z - a)^k f(z) \right).$$

If $k = 1$ (simple pole) we have

$$a_{-1} = \lim_{z \to a} (z - a) f(z).$$

The *residue* is given by

$$\text{Res}_{z=i\lambda} \frac{ze^{iaz}}{z^2 + \lambda^2} := \lim_{z \to i\lambda} \frac{z(z - i\lambda)e^{iaz}}{(z + i\lambda)(z - i\lambda)} = \frac{i\lambda e^{iai\lambda}}{2i\lambda} = \frac{1}{2} e^{-a\lambda}.$$

The *residue theorem* states that

$$\oint_C f(z) dz = 2\pi(a_{-1} + b_{-1} + c_{-1} + \cdots)$$

where $a_{-1}, b_{-1}, c_{-1}, \ldots$ are the residues inside C. Consequently,

$$\int_C \frac{ze^{iaz}}{z^2 + \lambda^2} dz = \pi i e^{-a\lambda}.$$

If $R \to \infty$, the second integral on the right hand side of (1) vanishes. Therefore

$$\int_{-\infty}^{+\infty} \frac{xe^{iax}}{x^2 + \lambda^2} dx = \pi i e^{-a\lambda}, \qquad \Im \int_{-\infty}^{+\infty} \frac{xe^{iax}}{x^2 + \lambda^2} dx = \pi e^{-a\lambda}.$$

Finally

$$\int_0^\infty \frac{x \sin(ax)}{x^2 + \lambda^2} dx = \frac{\pi}{2} e^{-a\lambda}.$$

Problem 4. (i) Calculate the Fourier transform $\hat{f}(\omega)$ of

$$f(t) = e^{-|t|}.$$

(ii) Calculate the inverse Fourier transform of $\hat{f}(\omega)$.

Solution 4. (i) From the definition of the Fourier transform

$$\hat{f}(\omega) := \int_{-\infty}^{+\infty} f(t) e^{i\omega t} dt$$

we obtain

$$\hat{f}(\omega) = \int_{-\infty}^0 e^t e^{i\omega t} dt + \int_0^\infty e^{-t} e^{i\omega t} dt \equiv \int_{-\infty}^0 e^{(i\omega+1)t} dt + \int_0^\infty e^{(i\omega-1)t} dt.$$

Integration yields

$$\hat{f}(\omega) = \frac{2}{1 + \omega^2}.$$

(ii) The inverse Fourier transform is given by

$$f(t) = \frac{1}{2\pi} \int_{-\infty}^{+\infty} \hat{f}(\omega) e^{-i\omega t} \, d\omega.$$

Therefore

$$f(t) = \frac{1}{\pi} \int_{-\infty}^{+\infty} \frac{1}{1+\omega^2} e^{-i\omega t} \, d\omega.$$

To calculate this integral we extend ω in the complex domain and consider

$$\frac{1}{\pi} \oint_C \frac{e^{-izt}}{1+z^2} \, dz.$$

Since the integrand has the two poles $z = \pm i$ and

$$\frac{1}{1+z^2} \equiv -\frac{1}{2i}\frac{1}{z+i} + \frac{1}{2i}\frac{1}{z-i}$$

we consider two paths C_1 and C_2. The path C_1 is the contour consisting of the line along the x-axis from $-R$ to R and the semicircle Γ above the x-axis having this line as diameter. The path C_2 is the contour consisting of the line along the x-axis from $-R$ to R and the semicircle Γ below the x-axis having this line as diameter. For the path C_1 we have

$$\frac{1}{\pi} \oint_{C_1} \frac{e^{-izt}}{1+z^2} \, dz \equiv -\frac{1}{2i\pi} \oint_{C_1} \frac{e^{-izt}}{z+i} \, dz + \frac{1}{2i\pi} \oint_{C_1} \frac{e^{-izt}}{z-i} \, dz.$$

The first integral on the right-hand side is equal to zero, since the pole $z = -i$ is not inside C_1. Consequently,

$$\frac{1}{\pi} \oint_{C_1} \frac{e^{-izt}}{1+z^2} \, dz = \frac{1}{2i\pi} \oint_{C_1} \frac{e^{-izt}}{z-i} \, dz.$$

Thus $f(t) = e^t$ for $t \in (-\infty, 0]$. Analogously, for the path C_2 we have

$$\frac{1}{\pi} \oint_{C_2} \frac{e^{-izt}}{1+z^2} \, dz = -\frac{1}{2i\pi} \oint_{C_2} \frac{e^{-izt}}{z+i} \, dz + \frac{1}{2i\pi} \oint_{C_2} \frac{e^{-izt}}{z-i} \, dz.$$

The second integral on the right-hand side is equal to zero, since the pole $z = i$ is not inside C_2. Consequently,

$$\frac{1}{\pi} \oint_{C_2} \frac{e^{-izt}}{1+z^2} \, dz = -\frac{1}{2i\pi} \oint_{C_2} \frac{e^{-izt}}{z+i} \, dz.$$

Thus $f(t) = e^{-t}$ for $t \in [0, \infty)$. Consequently,

$$f(t) = e^{-|t|}.$$

Problem 5. (i) Expand the function

$$\sqrt{(z-1)(z-2)} \qquad \text{for} \quad |z| > 2$$

into its *Laurent series*.
(ii) Expand the function

$$\ln\left(\frac{1}{1-z}\right) \qquad \text{for} \quad |z| > 1$$

into its Laurent series.

Solution 5. (i) For $|z| > 2$ we have

$$\sqrt{(z-1)(z-2)} = \pm z \left(1 - \frac{1}{z}\right)^{1/2} \left(1 - \frac{2}{z}\right)^{1/2}.$$

Applying the binomial expansions for the square root, we can write for $|z| > 2$

$$\pm z \left(1 - \binom{\frac{1}{2}}{1}\frac{1}{z} + \binom{\frac{1}{2}}{2}\frac{1}{z^2} - \cdots\right)\left(1 - \binom{\frac{1}{2}}{1}\frac{2}{z} + \binom{\frac{1}{2}}{2}\frac{2^2}{z^2} - \cdots\right)$$

$$= \pm \left(c_0 z - c_1 + \frac{c_2}{z} - \frac{c_3}{z^2} + \cdots\right)$$

where

$$c_n = \binom{\frac{1}{2}}{n} + 2\binom{\frac{1}{2}}{n-1}\binom{\frac{1}{2}}{1} + 2^2\binom{\frac{1}{2}}{n-2}\binom{\frac{1}{2}}{2} + \cdots + 2^n\binom{\frac{1}{2}}{n}.$$

(ii) The function does not have a Laurent expansion for $|z| > 1$, since the function is not single-valued there. If we put $z' = 1/z$, then we find

$$\ln\left(\frac{-z'}{1-z'}\right) = \ln(-z') + \ln\left(\frac{1}{1-z'}\right) = \ln\left(-\frac{1}{z}\right) + \frac{1}{z} + \frac{1}{2z^2} + \frac{1}{3z^3} + \cdots.$$

Problem 6. Assume that f is a meromorphic function in the finite plane $\mathbb{C}$ and has a finite number of poles $a_1, a_2, \ldots, a_m$ none of which is an integer. Assume, moreover, that

$$\lim_{z\to\infty} zf(z) = 0.$$

Then

$$\lim_{N\to\infty} \sum_{n=-N}^{n=N} f(n)$$

exists and

$$\lim_{N\to\infty} \sum_{n=-N}^{n=N} f(n) = -\sum_{k=1}^{m} \text{Res}(a_k; \pi f(z)\cot(\pi z))$$

where Res denotes the residue and $\cot(\pi z) \equiv \cos(\pi z)/\sin(\pi z)$. Calculate

$$\sum_{n=-\infty}^{\infty} \frac{1}{n^2+n+1}.$$

Solution 6.   The function

$$f(z) = \frac{1}{z^2+z+1}$$

is meromorphic in the finite plane $\mathbb{C}$ and has two poles, because

$$z^2+z+1 \equiv (z-a_1)(z-a_2)$$

with

$$a_1 = -\frac{1+i\sqrt{3}}{2}, \qquad a_2 = \bar{a}_1.$$

The residue of

$$\frac{\pi\cot\pi z}{z^2+z+1}$$

at $z = a_1$ is given by

$$\lim_{z\to a_1} \frac{(z-a_1)\pi\cot\pi z}{(z-a_1)(z-\bar{a}_1)} = \frac{\pi\cot(\pi a_1)}{a_1-\bar{a}_1} = -\frac{i}{\sqrt{3}}\pi\cot\left(\frac{\pi(1+i\sqrt{3})}{2}\right).$$

Similarly, the residue of

$$\frac{\pi\cot(\pi z)}{z^2+z+1}$$

at $z = \bar{a}_1$ is given by

$$\lim_{z\to\bar{a}_1} \frac{(z-\bar{a}_1)\pi\cot(\pi z)}{(z-a_1)(z-\bar{a}_1)} = \frac{\pi\cot(\pi\bar{a}_1)}{\bar{a}_1-a_1} = \frac{i}{\sqrt{3}}\pi\cot\left(\frac{\pi(1-i\sqrt{3})}{2}\right).$$

Therefore

$$\sum_{n=-\infty}^{\infty} \frac{1}{n^2+n+1} = -\frac{2\pi i}{\sqrt{3}}\tan\left(\frac{i\pi\sqrt{3}}{2}\right) = \frac{2\pi}{\sqrt{3}}\tanh\left(\frac{\pi\sqrt{3}}{2}\right)$$

where we have used the identity

$$\cot\left(\frac{\pi}{2} + \alpha\right) \equiv -\tan\alpha.$$

Problem 7. A function which is analytic everywhere in the finite complex plane (i.e. everywhere except at ∞) is called an *entire function*. For example, the functions $\sin z$, e^z and z^4 are entire functions. An entire function can be represented by a Taylor series which has an infinite radius of convergence. Conversely, if a power series has an infinite radius of convergence, it represents an entire function.

Solve the first order linear differential equation

$$\left(\frac{d}{dz} - \frac{\epsilon + \kappa^2}{z + \kappa} + \kappa\right) w(z) = 0 \tag{1}$$

where κ and ϵ are real constants. Impose the condition that w is an entire function. The differential equation (1) is the *Bargmann representation* of the *displaced harmonic oscillator*.

Solution 7. Method 1. Using the power series ansatz

$$w(z) = \sum_{n=0}^{\infty} b_n (z + \kappa)^{n+s}$$

yields

$$s = \epsilon + \kappa^2$$

and the recurrence relation for the expansion coefficients b_n's is given by

$$b_n = \frac{(-\kappa)}{n} b_{n-1}.$$

This recurrence relation can easily be solved

$$b_n = \frac{(-\kappa)^n}{\Gamma(n+1)} b_0$$

where Γ denotes the gamma function, i.e. $\Gamma(n+1) = n!$ with $n = 0, 1, 2, \ldots$. The series $\sum_n b_n z^n$ therefore converges for all z and

$$w(z) = b_0(z+\kappa)^{\epsilon+\kappa^2} \sum_{n=0}^{\infty} \frac{(-\kappa)^n}{\Gamma(n+1)}(z+\kappa)^n = b_0(z+\kappa)^{\epsilon+\kappa^2} e^{-\kappa(z+\kappa)}.$$

Imposing the condition that w is an entire function leads to

$$\epsilon + \kappa^2 = n$$

where $n = 0, 1, 2, \ldots$. In physics this is called a *quantization condition*.

Method 2. Equation (1) can be directly integrated since

$$\frac{dw}{dz} = \left(\frac{\epsilon + \kappa^2}{z + \kappa} - \kappa\right) w(z).$$

It follows that

$$\int \frac{dw}{w} = \int \left(\frac{\epsilon + \kappa^2}{z + \kappa} - \kappa\right) dz.$$

We find

$$w(z) = C(z + \kappa)^{\epsilon + \kappa^2} e^{-\kappa z}.$$

The requirement that w be entire implies that there be no branch-point singularities at $z = -\kappa$ and therefore

$$\epsilon + \kappa^2 = n.$$

Problem 8.   Assume $Ox_1x_2x_3$ is the system of rectangular coordinates whose axes Ox_1, Ox_2 coincide with the real and imaginary axes Ox, Oy of the complex plane $\mathbb{C}$. Assume, moreover, that the ray emanating from the north pole $N(0, 0, 1)$ of the unit sphere S^2

$$x_1^2 + x_2^2 + x_3^2 = 1 \tag{1}$$

and intersecting S^2 at $A(x_1, x_2, x_3)$ intersects $\mathbb{C}$ at the point z. The point $z = x + iy$ is then called the *stereographic projection* of $A(x_1, x_2, x_3)$ whereas A is called the spherical image of z. The stereographic projection is given by

$$x_1 = \frac{z + \bar{z}}{1 + |z|^2}, \qquad x_2 = \frac{z - \bar{z}}{i(1 + |z|^2)}, \qquad x_3 = \frac{|z|^2 - 1}{1 + |z|^2}$$

and

$$z = \frac{x_1 + ix_2}{1 - x_3}.$$

(i) Find the spherical images of $e^{i\alpha}$ where $\alpha \in \mathbb{R}$.
(ii) Let θ, ϕ be the geographical latitude and longitude of A respectively. Show that the stereographic projection of A has the representation

$$z = e^{i\phi} \tan\left(\frac{1}{4}\pi + \frac{1}{2}\theta\right). \tag{2}$$

(iii) The distance $\sigma(z_1, z_2)$ between two points on S^2 whose stereographic projections are z_1, z_2 is called the *spherical distance* or *chordal distance* between z_1 and z_2. Show that

$$\sigma(z_1, z_2) = \frac{2|z_1 - z_2|}{\sqrt{(1 + |z_1|^2)(1 + |z_2|^2)}} \tag{3}$$

is the Euclidean distance of the images of z_1 and z_2.

Solution 8. (i) Since $z = e^{i\alpha}$ we have $\bar{z} = e^{-i\alpha}$ and $z\bar{z} = 1$. Therefore

$$z + \bar{z} = 2\cos\alpha \qquad z - \bar{z} = 2i\sin\alpha.$$

It follows that

$$x_1 = \cos\alpha, \quad x_2 = \sin\alpha, \quad x_3 = 0.$$

(ii) Since spherical coordinates are given by

$$x_1(\theta, \phi) = \sin\theta\cos\phi, \quad x_2(\theta, \phi) = \sin\theta\sin\phi, \quad x_3(\theta, \phi) = \cos\theta$$

we obtain

$$z = \frac{x_1 + ix_2}{1 - x_3} = \frac{\sin\theta\cos\phi + i\sin\theta\sin\phi}{1 - \cos\theta} = \frac{e^{i\phi}\sin\theta}{1 - \cos\theta}.$$

Obviously, (2) follows.

(iii) Let $\mathbf{x}$ and $\mathbf{y}$ be two points on the unit sphere (1). Then

$$x_1^2 + x_2^2 + x_3^2 = 1, \qquad y_1^2 + y_2^2 + y_3^2 = 1. \tag{4}$$

It follows that

$$\|\mathbf{x} - \mathbf{y}\| = \sqrt{(x_1 - y_1)^2 + (x_2 - y_2)^2 + (x_3 - y_3)^2}$$
$$= \sqrt{2 - 2x_1y_1 - 2x_2y_2 - 2x_3y_3}$$

where we have used (4). Under the stereographic projection we have

$$z_1 \rightarrow (x_1, x_2, x_3), \qquad z_2 \rightarrow (y_1, y_2, y_3).$$

Now

$$1 + |z_1|^2 = \frac{2}{1 - x_3} \qquad 1 + |z_2|^2 = \frac{2}{1 - y_3} \tag{5}$$

and with $z_1 = (x_1 + ix_2)/(1 - x_3)$, $z_2 = (y_1 + iy_2)/(1 - y_3)$ we obtain

$$|z_1 - z_2| = \frac{\sqrt{2(1 - x_3)^2(1 - y_3^2) - 2(1 - x_3)(1 - y_3)(x_1y_1 + x_2y_2)}}{(1 - x_3)(1 - y_3)}.$$

Inserting this equation and (5) into (3) leads to

$$\sigma(z_1, z_2) = \|\mathbf{x} - \mathbf{y}\|.$$

The stereographic projection can be extended to arbitrary dimensions. Let S^n denote the unit n-sphere in $\mathbb{R}^{n+1}$ and let $\mathbf{q} = (0, \dots, 0, 1)$ denote the "north pole" of S^n. Let

$$\mathbf{f} : \mathbb{R}^n \rightarrow S^n$$

be the map which sends each $\mathbf{p} \in \mathbb{R}^n$ into the point different from $\mathbf{q}$ where the line through $(\mathbf{p}, 0) \in \mathbb{R}^{n+1}$ and $\mathbf{q}$ cuts S^n. Since

$$\boldsymbol{\alpha}(t) = t(\mathbf{p}, 0) + (1 - t)\mathbf{q} = (t\mathbf{p}, 1 - t)$$

is a parametrization of the line through $(\mathbf{p}, 0)$ and $\mathbf{q}$, and since

$$\|\boldsymbol{\alpha}(t)\| = 1$$

if and only if $t = 0$ or

$$t = \frac{2}{\|\mathbf{p}\|^2 + 1}$$

the map $\mathbf{f}$ is given by

$$\mathbf{f}(x_1, x_2, \ldots, x_n) = \frac{(2x_1, 2x_2, \ldots, 2x_n, x_1^2 + x_2^2 + \cdots + x_n^2 - 1)}{x_1^2 + x_2^2 + \cdots + x_n^2 + 1}.$$

The mapping $\mathbf{f}$ is a parametrized surface which maps $\mathbb{R}^n$ one-to-one onto $S^n \setminus \{\mathbf{q}\}$. The chart $\mathbf{f}^{-1}$ is called the stereographic projection from $S^n \setminus \{\mathbf{q}\}$ onto the equatorial hyperplane.

Problem 9.   A *Möbius transformation* is a function of the form

$$w(z) = \frac{az + b}{cz + d}$$

where a, b, c, d are complex numbers and $ad - bc \neq 0$. Find the Möbius transformation that carries the point 1 to i, the point 0 to 1 and the point i to ∞.

Solution 9.   Since 0 goes to 1, we must have $b/d = 1$. Thus

$$w(z) = \frac{az + b}{cz + b}.$$

Since i goes to ∞, $z = i$ must make the denominator zero. Hence we have $ci + b = 0$, $b = -ic$, and therefore

$$w(z) = \frac{az - ic}{cz - ic}.$$

Finally, since 1 goes to i

$$i = \frac{a - ic}{c - ic}.$$

Thus $a = c(1 + 2i)$. The coefficient c is at our disposal, with $c \neq 0$. Let $c = 1$. Then we arrive at

$$w(z) = \frac{(1 + 2i)z - i}{z - i}.$$

A Möbius transformation is a one-to-one map of the extended z plane to the extended w plane, with $w = 0$ when $az + b = 0$ and $w = \infty$ when $cz + d = 0$. The transformation is a composition of magnifications, rotations, translations, and a reciprocal transformation $w(z) = 1/z$.

Problem 10. Give the domain in the complex plane where the infinite series

$$\sum_{n=1}^{\infty} \frac{1}{n^2} \exp\left(\frac{nz}{z-2}\right) \tag{1}$$

converges.

Solution 10. Let

$$f(z) = \exp\left(\frac{z}{z-2}\right).$$

The series (1) can then be cast into the form

$$\sum_{n=1}^{\infty} \frac{1}{n^2}(f(z))^n.$$

Thus using the theory of power series, it converges if and only if $|f(z)| \le 1$. This inequality holds when

$$\Re\frac{z}{z-2} \le 0.$$

Thus we have to find the region which is sent into the closed left half-plane by the linear fractional map

$$h(z) = \frac{z}{z-2}.$$

The inverse of this map is the map

$$g(z) = \frac{2z}{z-1}.$$

Since $g(0) = 0$ and $g(\infty) = 2$, the image of the imaginary axis under g is a circle passing through the points 0 and 2. We can conclude that $|f(z)| \le 1$ if and only if $|z - 1| \le 1$ and $z \ne 2$, which is the region of convergence of the series (1).

Problem 11. Let

$$f(z) = \frac{P(z)}{Q(z)}$$

where P and Q are polynomials with real coefficients of degree m and n, respectively with $n \ge m + 2$. If the polynomial Q has simple zeros at the points $t_1, t_2, \ldots, t_r$ on the x-axis, then the *Cauchy principal value* (PV) of the integral is

$$PV \int_{-\infty}^{\infty} \frac{P(x)}{Q(x)} dx = 2\pi i \sum_{j=1}^{k} \text{Res} \left[\frac{P}{Q}, z_j \right] + \pi i \sum_{j=1}^{r} \text{Res} \left[\frac{P}{Q}, t_j \right]$$

where $z_1, z_2, \ldots, z_k$ are the poles of the function f that lie in the upper half-plane. Here Res denotes the residue. Evaluate

$$PV \int_{-\infty}^{\infty} \frac{x}{x^3 - 8} dx.$$

Solution 11.   Solving $z^3 = 8$ yields

$$z_1 = 2, \quad z_2 = -1 - i\sqrt{3}, \quad z_3 = -1 + i\sqrt{3}.$$

Thus z_1 lies on the real axis and z_3 lies in the upper half plane and $k = r = 1$. Next we compute the residues at $z_1 = 2$ and $z_3 = -1 + i\sqrt{3}$. We find

$$\text{Res}[f, 2] = \frac{1}{6}, \quad \text{Res}[f, -1 + i\sqrt{3}] = -\frac{i}{3(i + \sqrt{3})} = -\frac{1}{12} - \frac{i}{4\sqrt{3}}.$$

Thus

$$PV \int_{-\infty}^{\infty} \frac{x}{x^3 - 8} dx = 2\pi i \left(-\frac{1}{12} - \frac{i}{4\sqrt{3}} \right) + \frac{\pi i}{6} = \frac{\pi}{2\sqrt{3}}.$$

Problem 12.   The *Hilbert transform* of a function $f \in L_2(\mathbb{R})$ is defined as

$$H(f)(y) = \frac{1}{\pi} PV \int_{-\infty}^{\infty} \frac{f(x)}{x - y} dx$$

where PV stands for *principal value*. Calculate the Hilbert transform of $f(x) = 1/(1 + x^4)$.

Solution 12.   The Hilbert transform is given by

$$H(f)(y) = -\frac{1}{\sqrt{2}} \frac{y(1 + y^2)}{1 + y^4}.$$

Chapter 17

Special Functions

Legendre, Hermite, Laguerre and Chebyshev polynomials play an important role in mathematical physics, for example as an orthonormal basis in the Hilbert space of square integrable functions. Jacobi elliptic functions and Weierstrass elliptic functions play a role in the solution of nonlinear differential equations. The *Legendre polynomials* are given by the *Rodrigue's formula*

$$P_n(x) := \frac{1}{2^n n!} \frac{d^n}{dx^n} (x^2 - 1)^n, \qquad n = 0, 1, 2, \ldots$$

where $P_0(x) = 1$, $P_1(x) = x$, $P_2(x) = (3x^2 - 1)/2$. The *Hermite polynomials* are given by Rodrigue's formula

$$H_n(x) := (-1)^n e^{x^2} \frac{d^n}{dx^n} (e^{-x^2}), \qquad n = 0, 1, 2, \ldots$$

where $H_0(x) = 1$, $H_1(x) = 2x$, $H_2(x) = 4x^2 - 2$. The *Laguerre polynomials* are given by Rodrigue's formula

$$L_n(x) := e^x \frac{d^n}{dx^n} (x^n e^{-x}), \qquad n = 0, 1, 2, \ldots$$

where $L_0(x) = 1$, $L_1(x) = -x + 1$, $L_2(x) = x^2 - 4x + 2$. The *Chebyshev polynomials* are given by

$$T_n(x) = \cos(n \arccos(x)) = x^n - \binom{n}{2} x^{n-2}(1-x^2) + \binom{n}{4} x^{n-4}(1-x^2)^2 - \cdots$$

where $T_0(x) = 1$, $T_1(x) = x$, $T_2(x) = 2x^2 - 1$.

Problem 1.   (i) Consider the Legrendre polynomials given by the *formula of Rodrigues*. Let f be an analytic function. Show that the integral

$$I = \int_{-1}^{1} f(x) P_n(x) dx$$

can be brought into the form

$$I = \frac{(-1)^n}{2^n n!} \int_{-1}^{1} (x^2 - 1)^n \frac{d^n}{dx^n} f(x) dx. \tag{1}$$

(ii) Show that

$$\int_{-1}^{1} P_m(x) P_n(x) dx = 0, \qquad m \neq n. \tag{2}$$

Solution 1.   (i) Applying integration by parts to (1) we obtain

$$I = \frac{1}{2^n n!} \left| f(x) \frac{d^{n-1}}{dx^{n-1}} (x^2 - 1)^n \right|_{-1}^{+1} - \frac{1}{2^n n!} \int_{-1}^{1} \frac{df}{dx} \left(\frac{d^{n-1}}{dx^{n-1}} (x^2 - 1)^n dx \right). \tag{3}$$

Since $x^2 - 1$ is equal to zero at 1 and -1 we find that the first term on the right-hand side of (3) vanishes. Repeating this process of integration by parts we obtain (3).

(ii) Let $f(x) = P_m(x)$ with $m < n$. Then $d^n f/dx^n = 0$ and therefore (2) follows.

Problem 2.   A generating function for the *Legendre polynomials* $P_l(x)$ is given by

$$G(x, r) := \frac{1}{(1 - 2xr + r^2)^{1/2}} = \sum_{l=0}^{\infty} r^l P_l(x)$$

where $x = \cos\theta$ and $|r| < 1$. Prove that $P_l(x)$ satisfies the linear differential relation

$$x \frac{dP_l(x)}{dx} = \frac{P'_{l-1}(x)}{dx} + l P_l(x) \tag{1}$$

Solution 2.   Differentiating G with respect to r yields

$$\frac{(x - r)}{(1 - 2xr + r^2)^{3/2}} = \sum_{l=0}^{\infty} l r^{l-1} P_l(x). \tag{2}$$

Differentiating G with respect to x gives

$$\frac{r}{(1 - 2xr + r^2)^{3/2}} = \sum_{l=0}^{\infty} r^l \frac{dP'_l(x)}{dx}. \tag{3}$$

Eliminating $(1 - 2xr + r^2)^{-3/2}$ between (2) and (3) leads to

$$\sum_{l=0}^{\infty} (x - r)r^l \frac{dP_l(x)}{dx} = \sum_{l=0}^{\infty} lr^l P_l(x)$$

which, upon equating coefficients of equal powers of r, gives (1). The first three Legendre polynomials are given by

$$P_0(x) = 1, \qquad P_1(x) = x, \qquad P_2(x) = \frac{1}{2}(3x^2 - 1).$$

Problem 3. The *dominant tidal potential* at position (r, ϕ, λ) due to the moon or sun is given by

$$U(\mathbf{r}) = \frac{GM^* r^2}{r^{*3}} P_2^0(\cos \psi)$$

where M^* is the mass of the moon or sun located at (r^*, ϕ^*, λ^*). Moreover, ψ is the angle between mass M^* and the observation point at (r, ϕ, λ), where ϕ is the latitude and λ is the longitude. By the *spherical cosine theorem* we have

$$\cos \psi = \sin \phi \sin \phi^* + \cos \phi \cos \phi^* \cos(\lambda - \lambda^*).$$

The *Legendre polynomials* are defined by the Rodrigue's formula given above. The *associated Legendre polynomials* are defined as

$$P_n^m(x) := (1 - x^2)^{m/2} \frac{d^m}{dx^m} P_n(x) = \frac{(1 - x^2)^{m/2}}{2^n n!} \frac{d^{m+n}}{dx^{m+n}} (x^2 - 1)^n$$

with $P_n^0(x) = P_n(x)$ and $P_n^m = 0$ if $m > n$.
(i) Show that $U(\mathbf{r})$ can be written as

$$U(\mathbf{r}) = \frac{GM^* r^2}{r^{*3}} \left(P_2^0(\sin \phi) P_2^0(\sin \phi^*) + \frac{1}{3} P_2^1(\sin \phi) P_2^1(\sin \phi^*) \cos(\lambda - \lambda^*) \right.$$

$$\left. + \frac{1}{12} P_2^2(\sin \phi) P_2^2(\sin \phi^*) \cos(2(\lambda - \lambda^*)) \right).$$

(ii) Give an interpretation (maxima and nodes) of the terms in the parenthesis.

Solution 3. (i) Since

$$\cos^2 \psi = \sin^2 \phi \sin^2 \phi^* + \cos^2 \phi \cos^2 \phi^* \cos^2(\lambda - \lambda^*)$$

$$+ 2 \sin \phi \cos \phi \sin \phi^* \cos \phi^* \cos(\lambda - \lambda^*)$$

and $\cos^2 \alpha \equiv \dfrac{1}{2} + \dfrac{1}{2} \cos(2\alpha)$ we arrive at

$$\cos^2 \psi = \frac{3}{2} \sin^2 \phi \sin^2 \phi^* - \frac{1}{2}(\sin^2 \phi + \sin^2 \phi^*) + \frac{1}{2}$$
$$+ \frac{1}{2} \cos^2 \phi \cos^2 \phi^* \cos(2(\lambda - \lambda^*))$$
$$+ 2\sin\phi \cos\phi \sin\phi^* \cos\phi^* \cos(\lambda - \lambda^*).$$

Thus

$$P_2^0(\cos\psi) = \frac{1}{2}(3\cos^2 \psi - 1)$$
$$= \frac{9}{4} \sin^2 \phi \sin^2 \phi^* - \frac{3}{4}(\sin^2 \phi + \sin^2 \phi^*) - \frac{1}{4}$$
$$+ 3\sin\phi \cos\phi \sin\phi^* \cos\phi^* \cos(\lambda - \lambda^*)$$
$$+ \frac{3}{4} \cos^2 \phi \cos^2 \phi^* \cos(2(\lambda - \lambda^*)).$$

Since

$$P_2^0(\sin\phi)P_2^0(\sin\phi^*) = \frac{1}{4}(9\sin^2 \phi \sin^2 \phi^* - 3\sin^2 \phi - 3\sin^2 \phi^* + 1)$$
$$P_2^1(\sin\phi)P_2^1(\sin\phi^*) = 9\sin\phi \cos\phi \sin\phi^* \cos\phi^*$$
$$P_2^2(\sin\phi)P_2^2(\sin\phi^*) = 9\cos^2 \phi \cos^2 \phi^*$$

we find the expression for $U(\mathbf{r})$.

(ii) The first term in the parenthesis has a maximum at the poles and 2 nodes in the latitude but no longitude dependence. Thus it represents long periodic zonal tides. The second term has a maximum at $\pm 45°$ latitude, one nodal line along the equator and two nodal longitudes. Thus it represents diurnal tesseral tides. The third term has a maximum at the equator and no nodes in the latitude other than those at the poles, but 4 nodal longitudes. Thus it represents semi-diurnal sectorial tides. Since the location of M^*, i.e., (r^*, ϕ^*, λ^*) are time dependent, the terms

$$P_2^m(\sin\phi^*) \cos(m(\lambda - \lambda^*))$$

for the sun or moon can be expanded in trigonometric series in t, m and λ.

Problem 4.   A generating function $F(x, t)$ of the *Hermite polynomials* $H_n(x)$ is given by

$$F(x, t) := e^{x^2 - (t-x)^2} = \sum_{k=0}^{\infty} H_k(x)\frac{t^k}{k!}$$

where $n = 0, 1, 2, \ldots$ and $t, x \in \mathbb{R}$.

(i) Express $H_n(x)$ as a contour integral.

(ii) Prove that $H_n(x)$ satisfies *Hermite's differential equation*

$$\frac{d^2 H_n}{dx^2} - 2x\frac{dH_n}{dx} + 2nH_n = 0.$$

(iii) Show that

$$\frac{dH_n}{dx} = 2nH_{n-1}, \quad n = 1, 2, \ldots.$$

Solution 4.   (i) We know that

$$\oint \frac{1}{z^n}dz = \begin{cases} 2\pi i & \text{if } n = 1 \\ 0 & \text{otherwise} \end{cases} \tag{1}$$

where the integration is performed over a closed contour in the complex z-plane enclosing the origin and $n \in \mathbb{Z}$. Dividing the generating equation $F(x, t)$ by t^{n+1} and integrating over a closed contour in the complex t-plane enclosing the origin, we obtain

$$H_n(x) = \frac{n!}{2\pi i} \oint \frac{\exp(x^2 - (t - x)^2)}{t^{n+1}}dt \tag{2}$$

where we have used (1).

(ii) One forms the quantity

$$\frac{\partial^2 F}{\partial x^2} + 2t\frac{\partial F}{\partial t} - 2x\frac{\partial F}{\partial x}$$

and verifies that it is identically zero. Using the expansion for F in terms of H_n, this identity takes the form

$$\sum_{n=0}^{\infty} \frac{(H_n''(x) - 2xH_n'(x) + 2nH_n(x))t^n}{n!} = 0$$

for all t, where $H_n' \equiv dH_n/dx$. It follows that

$$H_n''(x) - 2xH_n'(x) + 2nH_n(x) = 0.$$

(iii) Differentiating the integral representation for $H_n(x)$ given by (2) with respect to x we obtain

$$\frac{dH_n}{dx} = \frac{2n((n-1)!)}{2\pi i} \oint \frac{\exp(x^2 - (t - x)^2)}{t^n}dt.$$

Taking into account (2) we find that

$$\frac{dH_n}{dx} = 2nH_{n-1}.$$

Problem 5.   Let $L_n(x)$, $H_n(x)$ be the Laguerre and Hermite polynomials, where $n = 0, 1, \ldots$. Let

$$L_n^{(\alpha)}(x) := \frac{x^{-\alpha}e^x}{n!} \frac{d^n}{dx^n}(e^{-x}x^{n+\alpha})$$

be the associated Laguerre polynomials with $\alpha > -1$ and $n = 0, 1, \ldots$. The Laguerre polynomials are recovered by setting $\alpha = 0$. We have

$$H_{2n}(x) = (-4)^n n! L_n^{(-1/2)}(x^2) \tag{1}$$

and the following addition formula for the associated Laguerre polynomials $L_n^\alpha(x)$

$$L_n^{(\alpha+\beta+1)}(x+y) = \sum_{k=0}^{n} L_{n-k}^{(\alpha)}(x) L_k^{(\beta)}(y). \tag{2}$$

(i) Find a new sum rule by inserting (1) into (2).
(ii) Consider the *sum rule*

$$\frac{1}{n!2^n} H_n(\sqrt{2}x) H_n(\sqrt{2}y) = \sum_{k=1}^{n} (-1)^k L_{n-k}^{(-1/2)}((x+y)^2) L_k^{(-1/2)}((x-y)^2). \tag{3}$$

Insert (1) into (3) to find a sum rule for Hermite polynomials.

Solution 5.   (i) Setting $\alpha = \beta = -1/2$, $x \to x^2$, $y \to y^2$ we obtain

$$(-4)^n L_n(x^2 + y^2) = \sum_{k=0}^{n} \frac{1}{(n-k)!k!} H_{2n-2k}(x) H_{2k}(y).$$

(ii) Setting $x + y \to x$ and $x - y \to y$ we obtain

$$\frac{(-2)^n}{n!} H_n\left(\frac{x+y}{\sqrt{2}}\right) H_n\left(\frac{x-y}{\sqrt{2}}\right) = \sum_{k=0}^{n} \frac{(-1)^k}{(n-k)!k!} H_{2n-2k}(x) H_{2k}(y).$$

Problem 6.   Consider the Hilbert space $L_2[0, \infty)$. An orthonormal basis $\{\phi_n(x) : n = 0, 1, 2, \ldots\}$ is given by

$$\phi_n(x) = e^{-x/2} L_n(x)$$

where the Laguerre polynomials are defined by

$$L_n(x) := \frac{1}{n!} e^x \frac{d^n}{dx^n}(x^n e^{-x}).$$

(i) Find ϕ_0, ϕ_1 and ϕ_2.

(ii) Calculate the scalar product $\langle \phi_2(x) | \phi_2(x) \rangle$. We note that

$$\int_0^\infty x^n e^{-x} dx = n! \,.$$

(iii) Consider the function $f : [0, \infty) \to \mathbb{R}$

$$f(x) = \begin{cases} 1 & \text{for} \quad 0 \le x \le 1 \\ 0 & \text{for} \quad x > 1 \end{cases}$$

Calculate the coefficients $c_0 = \langle f, \phi_0 \rangle$, $c_1 = \langle f, \phi_1 \rangle$ of the expansion

$$f(x) = \sum_{n=0}^\infty \langle f, \phi_n \rangle \phi_n(x)$$

i.e. $c_n = \langle f, \phi_n \rangle$.

Solution 6. (i) We have

$$\phi_0(x) = e^{-x/2}, \qquad \phi_1(x) = (-x+1)e^{-x/2}, \qquad \phi_2(x) = \frac{1}{2}(x^2 - 4x + 2)e^{-x/2}.$$

(ii) For the scalar product we find

$$\begin{aligned}
\langle \phi_2(x) | \phi_2(x) \rangle &= \frac{1}{4} \int_0^\infty (x^2 - 4x + 2)^2 e^{-x} dx \\
&= \frac{1}{4} \int_0^\infty (x^4 - 8x^3 + 20x^2 - 16x + 4)e^{-x} dx \\
&= \frac{1}{4}(4! - 8 \cdot 3! + 20 \cdot 2! - 16 \cdot 1! + 4) \\
&= 1 \,.
\end{aligned}$$

This is obvious since the functions ϕ_n form an orthonormal basis in the Hilbert space $L_2[0, \infty)$.

(iii) We have

$$c_0 = \int_0^\infty f(x)\phi_0(x)dx = \int_0^1 e^{-x/2} dx = -2e^{-1/2} + 2$$

$$c_1 = \int_0^\infty f(x)\phi_1(x)dx = \int_0^1 e^{-x/2}(-x+1)dx = 4e^{-1/2} - 2 \,.$$

Problem 7. Consider

$$T_{n+1}(x) - 2xT_n(x) + T_{n-1}(x) = 0 \tag{1}$$

where $n = 1, 2, \ldots,$

$$T_0(x) = 1, \qquad T_1(x) = x \tag{2}$$

and x is a fixed parameter. Solve the linear difference equation (1) where the initial values are given by (2). The quantities $T_n(x)$ are called *Chebyshev polynomials* of the first kind.

Solution 7.   Since (1) is a linear difference equation with constant coefficients it can be solved with the ansatz

$$T_n(x) := A(r(x))^n$$

where A is a constant. Inserting this ansatz into (1) yields

$$r^{n+1} - 2xr^n + r^{n-1} = 0$$

or $r^2 - 2xr + 1 = 0$. The solution to this quadratic equation is given by

$$r_{1,2}(x) = x \pm \sqrt{x^2 - 1}.$$

Thus the solution of the difference equation (1) takes the form

$$T_n(x) = A(x + \sqrt{x^2 - 1})^n + B(x - \sqrt{x^2 - 1})^n$$

where A and B are the constants of integration and $n = 0, 1, 2, \ldots$. Imposing the initial conditions (2) yields

$$T_n(x) = \frac{1}{2}(x + \sqrt{x^2 - 1})^n + \frac{1}{2}(x - \sqrt{x^2 - 1})^n.$$

The solution can also be written as

$$T_n(x) = \cos(n \arccos x), \quad n = 0, 1, 2, \ldots.$$

Problem 8.   (i) Let n be a positive integer. Let $f : [a, b] \to \mathbb{R}$ be a continuous function. The *Bernstein polynomials* of degree n associated with the continuous function f are given by

$$B_n(f, x) := \frac{1}{(b-a)^n} \sum_{j=0}^{n} \binom{n}{j} (x-a)^j (b-x)^{n-j} f(x_j)$$

where

$$x_j = a + j \frac{b-a}{n}, \qquad j = 0, 1, \ldots, n.$$

Consider the function $f : [0, 1] \to \mathbb{R}$ given by $f(x) = \sin(4x)$. Show that $B_2(f, x)$ is not a "good approximation" for f.

(ii) The *Bernstein basis polynomials* are defined as

$$B_{n,j}(x) := \binom{n}{j} x^j (1-x)^{n-j}, \quad j = 0, 1, 2, \ldots, n, \quad x \in [0, 1].$$

Show that

$$\sum_{j=0}^{n} B_{n,j}(x) = 1.$$

Show that $B_{n,j}$ satisfies the recursion relations

$$B_{n,j}(x) = (1-x)B_{n-1,j}(x) + xB_{n,j-1}(x), \quad j = 1, 2, \ldots, n-1$$
$$B_{n,0}(x) = (1-x)B_{n-1,0}(x)$$
$$B_{n,n}(x) = xB_{n-1,n-1}(x).$$

Solution 8. (i) For $x = \pi/8$ we have $\sin(\pi/2) = 1$. Using $\sin(0) = 0$ and

$$\sin(2) = 2\sin(1)\cos(1), \quad \sin(4) = 4\sin(1)\cos(1) - 8\sin^3(1)\cos(1)$$

we have

$$B_2(f(x = \pi/8), x = \pi/8) = 4x\sin(1)\cos(1)(1 - 2x\sin^2(1)).$$

Thus $1 - B_2(f(x = \pi/8), x = \pi/8) \approx 0.65$.

(ii) We have

$$\sum_{j=0}^{n} B_{n,j}(x) = \sum_{j=0}^{n} \binom{n}{j} x^j (1-x)^{n-j} = ((1-x) + x)^n = 1.$$

To prove the recursion relation, we note that

$$(1-x)B_{n-1,j}(x) + xB_{n,j-1}(x)$$

$$= \frac{(n-1)!}{j!(n-1-j)!} x^j (1-x)^{n-j-1} (1-x)$$

$$+ \frac{(n-1)!}{(j-1)!(n-1-j+1)!} x^{j-1}(1-x)^{n-j} x$$

$$= x^j (1-x)^{n-j} \left(\frac{(n-1)!}{j!(n-1-j)!} + \frac{(n-1)!}{(j-1)!(n-j)!} \right)$$

$$= x^j (1-x)^{n-j} \frac{n!}{j!(n-j)!} \left(\frac{n-j}{n} + \frac{j}{n} \right)$$

$$= B_{n,j}(x).$$

Problem 9. Let $a > b > 0$. Then the equation of the *ellipse* in parametric form is given by

$$x(\phi) = a\sin\phi, \quad y(\phi) = b\cos\phi$$

where $0 \leq \phi < 2\pi$. The *metric tensor field* of the two-dimensional Euclidean space is given by

$$g = dx \otimes dx + dy \otimes dy. \tag{1}$$

Calculate the arc length of an *ellipse*.

Solution 9.   Since

$$dx = a\cos(\phi)d\phi, \qquad dy = -b\sin(\phi)d\phi$$

we obtain

$$dx \otimes dx = a^2\cos^2(\phi)d\phi \otimes d\phi, \qquad dy \otimes dy = b^2\sin^2(\phi)d\phi \otimes d\phi.$$

Consequently, we obtain from (1)

$$g = (a^2\cos^2\phi + b^2\sin^2\phi)d\phi \otimes d\phi.$$

Therefore the *line element* is given by

$$\left(\frac{ds}{d\phi}\right)^2 = a^2\cos^2\phi + b^2\sin^2\phi.$$

Using the identity $\sin^2\phi + \cos^2\phi \equiv 1$ we arrive at

$$\int_0^{2\pi} ds = \int_0^{2\pi} \sqrt{a^2 - (a^2 - b^2)\sin^2\phi}\, d\phi$$

or

$$\int_0^{2\pi} ds = a\int_0^{2\pi} \sqrt{1 - \frac{a^2 - b^2}{a^2}\sin^2\phi}\, d\phi.$$

This is a *complete elliptic integral of the second kind*, where

$$k^2 = \frac{a^2 - b^2}{a^2} =: e^2.$$

Here e is the *eccentricity* of the ellipse. We find

$$\int_0^{2\pi} ds = 4aE(k, \pi/2).$$

The complete elliptic integral of the second kind can be given as

$$E(k, \pi/2) = \int_0^{\pi/2} \sqrt{1 - k^2\sin^2\phi}\, d\phi.$$

Thus

$$E(k, \pi/2) = \frac{\pi}{2}\left(1 - \left(\frac{1}{2}\right)^2 k^2 - \left(\frac{1\cdot 3}{2\cdot 4}\right)^2 \frac{k^4}{3} - \left(\frac{1\cdot 3\cdot 5}{2\cdot 4\cdot 6}\right)^2 \frac{k^6}{5} - \cdots\right)$$

where $|k^2| < 1$. We have used the *Taylor series expansion* around $x = 0$ of the function

$$f(x) = \sqrt{1 - x^2}$$

with $x^2 < 1$, i.e.

$$\sqrt{1 - x^2} = 1 - \frac{1}{2}x^2 - \frac{1}{2 \cdot 4}x^4 - \frac{1 \cdot 3}{2 \cdot 4 \cdot 6}x^6 - \cdots, \qquad -1 < x \leq 1.$$

Problem 10. (i) Let

$$\operatorname{sn}^{-1}(s, k) := \int_0^{\arcsin(s)} \frac{dx}{\sqrt{1 - k^2 \sin^2 x}}$$

where $k \in [0, 1]$. Calculate

$$\frac{d\operatorname{sn}^{-1}(s, k)}{ds}.$$

(ii) Let

$$\operatorname{cn}^{-1}(s, k) := \int_0^{\arccos(s)} \frac{dx}{\sqrt{1 - k^2 \sin^2 x}}.$$

Calculate

$$\frac{d\operatorname{cn}^{-1}(s, k)}{ds}.$$

Solution 10. We use the *Leibniz rule* for differentiation of integrals

$$\frac{d}{ds} \int_{f(s)}^{g(s)} F(x, s)dx = \int_{f(s)}^{g(s)} \frac{\partial F}{\partial s}dx + F(g(s), s)\frac{dg}{ds} - F(f(s), s)\frac{df}{ds}. \qquad (1)$$

If F is independent of s, then (1) simplifies to

$$\frac{d}{ds} \int_{f(s)}^{g(s)} F(x)dx = F(g(s))\frac{dg}{ds} - F(f(s))\frac{df}{ds}.$$

(i) Applying this rule we obtain

$$\frac{d\operatorname{sn}^{-1}(s, k)}{ds} = \frac{1}{\sqrt{1 - k^2(\sin(\arcsin s))^2}} \frac{d}{ds} \arcsin(s).$$

Since $\sin(\arcsin(s)) = s$ and

$$\frac{d}{ds} \arcsin s = \frac{1}{\sqrt{1 - s^2}}$$

we obtain

$$\frac{d\mathrm{sn}^{-1}(s,k)}{ds} = \frac{1}{\sqrt{1-k^2s^2}}\frac{1}{\sqrt{1-s^2}}.$$

(ii) We find

$$\frac{d\mathrm{cn}^{-1}(s,k)}{ds} = \frac{1}{\sqrt{1-k^2(\sin(\arccos s))^2}}\frac{d}{ds}\arccos(s).$$

Since

$$\sin(\arccos(s)) = \sqrt{1-s^2}$$

and

$$\frac{d}{ds}\arccos(s) = -\frac{1}{\sqrt{1-s^2}}$$

we obtain

$$\frac{d\mathrm{cn}^{-1}(s,k)}{ds} = -\frac{1}{\sqrt{1-k^2(1-s^2)}}\frac{1}{\sqrt{1-s^2}}.$$

Problem 11.   Consider the function

$$\sigma(z;\omega_1,\omega_2) = z\prod_k{}' \left(1 - \frac{z}{\Omega_k}\right)\exp\left(\frac{z}{\Omega_k} + \frac{1}{2}\left(\frac{z}{\Omega_k}\right)^2\right)$$

with

$$\Omega_k = m_k\omega_1 + n_k\omega_2$$

where m_k, n_k is the sequence of all pairs of integers. The prime after the product sign indicates that the pair $(0,0)$ should be omitted. ω_1 and ω_2 are two complex numbers with

$$\Im(\omega_1/\omega_2) \neq 0.$$

In the following the dependence on ω_1 and ω_2 will be omitted. The logarithmic derivative $\sigma'(z)/\sigma(z)$ where $\sigma'(z) \equiv d\sigma/dz$ is the meromorphic function $\zeta(z)$ of Weierstrass. The function

$$\wp(z) = -\zeta'(z)$$

is a meromorphic, doubly periodic (or elliptic) function with periods ω_1, ω_2 whose only singularities are double poles $m\omega_1 + n\omega_2$. We find

$$\wp(z;\omega_1,\omega_2) = \frac{1}{z^2} + \sum_k{}' \left(\frac{1}{(z-\Omega_k)^2} - \frac{1}{(\Omega_k)^2}\right).$$

The function $\wp$ is called *Weierstrass function* $\wp$.

(i) Show that

$$\sigma(z) = z + c_5 z^5 + c_7 z^7 + \cdots . \tag{1}$$

(ii) Show that

$$\frac{\sigma(2u)}{\sigma^4(u)} = -\wp'(u) \tag{2a}$$

$$2\zeta(2u) - 4\zeta(u) = \frac{\wp''(u)}{\wp'(u)}. \tag{2b}$$

Solution 11. (i) Grouping the factors corresponding to the pairs (m, n), $(-m, -n)$ we find

$$\sigma(z) = z \prod_k{}' \left(1 - \frac{z}{\Omega_k^2}\right) \exp(z^2/\Omega_k^2).$$

Since every factor has the form $1 - z^4/\Omega_k^2$ we find (1).
(ii) The function

$$\frac{\sigma(2u)}{\sigma^4(u)}$$

is doubly periodic and has poles of order three at $m\omega_1 + n\omega_2$. Furthermore the function is equal to

$$2\frac{1}{u^3} + Au + \cdots$$

near the origin. Consequently,

$$\wp'(u) + \frac{\sigma(2u)}{\sigma^4(u)} = 0$$

since the left hand side has only removable singularities.
(iii) Calculating the *logarithmic derivatives*

$$(\ln f)' = \frac{f'}{f}$$

of both sides of (2a) and $\zeta(z) = \sigma'(z)/\sigma(z)$ we find (2b).

Problem 12. Let $a(q)$ be a meromorphic function. Let $r, k, l = 1, 2, \ldots, N$. Assume that a satisfies the following equation

$$\frac{a(q_k - q_r)a'(q_r - q_l) - a'(q_k - q_r)a(q_r - q_l)}{a(q_k - q_l)} = g(q_k - q_r) - g(q_r - q_l)$$

where a' denotes differentiation with respect to the arguments and $k \neq l$. Let

$$U(q) = a^2(q).$$

Show that

$$[U(x)U'(y) - U'(x)U(y)] + [U(y)U'(z) - U'(y)U(z)]$$
$$+ [U(z)U'(x) - U'(z)U(x)] = 0$$

where

$$x + y + z = (q_k - q_r) + (q_r - q_l) + (q_l - q_k) = 0.$$

Solution 12.   The permutations of (k, r, l) are given by

$$(k, r, l), \quad (k, l, r), \quad (r, k, l), \quad (r, l, k), \quad (l, k, r), \quad (l, r, k).$$

Adding all equations (1) corresponding to all permutations of (k, r, l) we find

$$\frac{a(q_r - q_l)a'(q_l - q_k) - a'(q_r - q_l)a(q_l - q_k)}{a(q_k - q_r)}$$
$$+ \frac{a(q_l - q_k)a'(q_k - q_r) - a'(q_l - q_k)a(q_k - q_r)}{a(q_r - q_l)}$$
$$+ \frac{a(q_k - q_r)a'(q_r - q_l) - a'(q_k - q_r)a(q_r - q_l)}{a(q_l - q_k)} = 0.$$

Therefore the equation for U follows. This equation is the *addition formula* for the Weierstrass $\wp$ function.

Chapter 18

Inequalities

In a normed vector space V, for example $\mathbb{C}^n$, the *triangular inequality* is

$$\|\mathbf{v} + \mathbf{w}\| \leq \|\mathbf{v}\| + \|\mathbf{w}\| \quad \text{for all} \quad \mathbf{v}, \mathbf{w} \in V.$$

Besides the triangular inequality the Cauchy-Schwarz, Hölder's, Bessel's, Parseval's, Gibb's and Jensen's inequalities play a central role in mathematical and theoretical physics.

Problem 1. (i) Let $a, b \in \mathbb{R}$. Show that

$$a^2 + b^2 \geq 2ab.$$

(ii) Let a, b be two non-negative numbers. Show that

$$\frac{1}{2}(a + b) \geq \sqrt{ab}.$$

Solution 1. (i) Since $(a - b)^2 \geq 0$ we find

$$a^2 + b^2 - 2ab \geq 0.$$

It follows that

$$a^2 + b^2 \geq 2ab.$$

(ii) Since

$$(\sqrt{a} - \sqrt{b})^2 \geq 0$$

203

we have

$$a + b - 2\sqrt{ab} \geq 0.$$

Therefore

$$a + b \geq 2\sqrt{ab}.$$

Problem 2.   Consider the differentiable function $f : [0, \infty) \to \mathbb{R}$

$$f(x) = \frac{x}{1 + x}.$$

Let $a, b \in \mathbb{R}^+$. Show that

$$f(|a + b|) \leq f(|a| + |b|). \tag{1}$$

Solution 2.   Obviously we have (triangular inequality)

$$|a + b| \leq |a| + |b|.$$

Differentiation of f gives

$$\frac{df}{dx} = \frac{1}{(1 + x)^2}$$

which is positive. Thus the function f is monotone increasing. Consequently inequality (1) follows.

Problem 3.   Let $a, b \in \mathbb{R}$. Show that

$$\frac{|a + b|}{1 + |a + b|} \leq \frac{|a|}{1 + |a|} + \frac{|b|}{1 + |b|}.$$

Solution 3.   From the triangular inequality

$$|a + b| \leq |a| + |b|$$

we obtain

$$\frac{|a + b|}{1 + |a + b|} \leq \frac{|a| + |b|}{1 + |a| + |b|}$$

$$= \frac{|a|}{1 + |a| + |b|} + \frac{|b|}{1 + |a| + |b|}$$

$$\leq \frac{|a|}{1 + |a|} + \frac{|b|}{1 + |b|}.$$

Problem 4. Find all $x \in \mathbb{R}$ such that

$$x^2 - x - 6 \leq 0.$$

Solution 4. We have

$$(x + 2)(x - 3) \leq 0.$$

Thus if $x \in [-2, 3]$ the inequality is satisfied. For all other x the inequality is not satisfied.

Problem 5. Let $a_1, a_2, \ldots, a_n, b_1, b_2, \ldots, b_n \in \mathbb{R}$.
(i) Show that

$$(a_1 b_1 + a_2 b_2 + \cdots + a_n b_n)^2 \leq (a_1^2 + a_2^2 + \cdots + a_n^2)(b_1^2 + b_2^2 + \cdots + b_n^2). \quad (1)$$

(ii) Show that

$$\sqrt{(a_1 + b_1)^2 + \cdots + (a_n + b_n)^2} \leq \sqrt{a_1^2 + \cdots + a_n^2} + \sqrt{b_1^2 + \cdots + b_n^2}. \quad (2)$$

Solution 5. (i) If $a_1 = a_2 = \cdots = a_n = 0$, then inequality (1) is satisfied. Let us now assume that at least one of the a_j's is nonzero. Let ϵ be an arbitrary real number. Then we obviously have

$$(a_1 \epsilon + b_1)^2 + (a_2 \epsilon + b_2)^2 + \cdots + (a_n \epsilon + b_n)^2 \geq 0. \quad (3)$$

Let

$$A^2 := a_1^2 + a_2^2 + \cdots + a_n^2, \qquad B^2 := b_1^2 + b_2^2 + \cdots + b_n^2$$

and

$$C := a_1 b_1 + a_2 b_2 + \cdots + a_n b_n.$$

Then from (3) we have

$$A^2 \epsilon^2 + 2C\epsilon + B^2 \geq 0.$$

The left-hand side of this inequality is a smooth function of ϵ, say

$$f(\epsilon) := A^2 \epsilon^2 + 2C\epsilon + B^2.$$

Since

$$\frac{df}{d\epsilon} = 2A^2 \epsilon + 2C, \qquad \frac{d^2 f}{d\epsilon^2} = 2A^2 > 0$$

we find that f has a minimum at

$$\epsilon = -\frac{C}{A^2}.$$

Inserting ϵ into the left-hand side of the inequality yields

$$\frac{A^2C^2}{A^4} - \frac{2C^2}{A^2} + B^2 \geq 0$$

or

$$A^2 B^2 \geq C^2.$$

This is inequality (1).
(ii) Obviously

$$0 \leq \sum_{j=1}^{n}(a_j + b_j)^2 \equiv \sum_{j=1}^{n}a_j^2 + \sum_{j=1}^{n}b_j^2 + 2\sum_{j=1}^{n}a_jb_j.$$

From (1) we have

$$a_1b_1 + a_2b_2 + \cdots + a_nb_n \equiv \sum_{j=1}^{n}a_jb_j \leq \sqrt{\sum_{j=1}^{n}a_j^2}\sqrt{\sum_{j=1}^{n}b_j^2}.$$

Therefore we obtain

$$\sum_{j=1}^{n}(a_j + b_j)^2 \leq \sum_{j=1}^{n}a_j^2 + \sum_{j=1}^{n}b_j^2 + 2\sqrt{\sum_{j=1}^{n}a_j^2}\sqrt{\sum_{j=1}^{n}b_j^2}.$$

Taking the square root on both sides leads to (2).

Problem 6.   Let $x \geq 0$ and

$$f(x) = x \ln x - x + 1. \tag{1}$$

Show that

$$f(x) \geq 0. \tag{2}$$

From L'Hospital's rule we find that $f(0) = 1$.

Solution 6.   From (1) we obtain

$$\frac{df}{dx} = \ln x \tag{3}$$

and

$$\frac{d^2 f}{dx^2} = \frac{1}{x}. \tag{4}$$

From (3) and (4) we find that the function f has one minimum at $x = 1$ for $x \geq 0$, since $\ln 1 = 0$ and

$$\left.\frac{d^2 f}{dx^2}\right|_{x=1} = 1 > 0.$$

It follows that $f(1) = 0$. Therefore inequality (2) follows. The inequality (2) is called the *Gibbs' inequality* and plays an important rôle in statistical physics.

Problem 7. Suppose that $f : \mathbb{R} \to \mathbb{R}$ is twice differentiable with

$$f''(x) \geq 0$$

for all x. Prove that for all a and b, $a < b$,

$$f\left(\frac{a+b}{2}\right) \leq \frac{f(a) + f(b)}{2}. \tag{1}$$

Solution 7. By the *mean-value theorem* there is a number x_1 in the interval $(a, \frac{1}{2}(a+b))$ such that

$$\frac{f(\frac{1}{2}(a+b)) - f(a)}{\frac{1}{2}(a+b) - a} = f'(x_1)$$

and a number x_2 in $(\frac{1}{2}(a+b), b)$ such that

$$\frac{f(b) - f(\frac{1}{2}(a+b))}{b - \frac{1}{2}(a+b)} = f'(x_2).$$

However $f''(x) \geq 0$ for all x in (x_1, x_2), so f' is a nondecreasing function. Thus

$$f'(x_2) \geq f'(x_1)$$

or equivalently,

$$\frac{f(b) - f(\frac{1}{2}(a+b))}{b - a} \geq \frac{f(\frac{1}{2}(a+b)) - f(a)}{b - a}.$$

Thus (1) follows.

Problem 8. Let $a_j > 0$ for $j = 1, 2, \ldots, n$. The *arithmetic mean* of $a_1, a_2, \ldots, a_n$ is the number

$$\frac{a_1 + a_2 + \cdots + a_n}{n}$$

and the *geometric mean* of $a_1, a_2, \ldots, a_n$ is the number

$$(a_1 a_2 \cdots a_n)^{1/n}.$$

(i) Show that

$$(a_1 a_2 \cdots a_n)^{1/n} \leq \frac{a_1 + a_2 + \cdots + a_n}{n}. \tag{1}$$

(ii) Consider a rectangular parallelepiped. Let the lengths of the three adjacent sides be a, b and c. Let A and V be the surface area and the volume, respectively. Prove that

$$A \geq 6V^{2/3}. \tag{2}$$

Solution 8.   (i) Let $n = 2$. Then we have to prove that

$$(a_1 a_2)^{1/2} \leq \frac{a_1 + a_2}{2}.$$

This inequality was proved in problem 1. The proof for larger values of n can be handled by mathematical induction.
(ii) Obviously

$$A = 2(ab + bc + ca), \qquad V = abc.$$

Thus $V^2 = a^2 b^2 c^2 = (ab)(bc)(ca)$. Using the *arithmetic mean geometric mean inequality* we obtain

$$V^2 \leq \left(\frac{ab + bc + ca}{3} \right)^3 = \left(\frac{2(ab + bc + ca)}{6} \right)^3 = \left(\frac{A}{6} \right)^3.$$

Thus inequality (2) follows. If $ab = bc = ca$ or equivalently $a = b = c$ we obtain

$$6V^{2/3} = A.$$

Problem 9.   Let $x \in \mathbb{R}$ and $x > 0$. Show that

$$\sqrt[e]{e} \geq \sqrt[x]{x}. \tag{1}$$

Solution 9.   Since

$$f(x) = \sqrt[x]{x} \equiv \exp\left(\frac{1}{x} \ln x \right), \qquad x > 0$$

we find

$$\frac{df}{dx} = \frac{1 - \ln x}{x^2} \exp\left(\frac{1}{x} \ln x \right).$$

The necessary condition for an extremum is $df/dx = 0$. It follows that

$$1 - \ln(x) = 0 \Rightarrow x = e \quad \text{(critical point)}.$$

Therefore

$$f(e) = e^{1/e} \approx 1.44466786101\ldots.$$

The second derivative is

$$\frac{d^2 f}{dx^2} = \left(\left(\frac{1 - \ln x}{x^2} \right)^2 + \frac{-x - (1 - \ln x)2x}{x^4} \right) \exp\left(\frac{1}{x} \ln x \right).$$

Setting $x = e$ and using $\ln e = 1$ yields

$$\frac{d^2 f}{dx^2} \bigg|_{x=e} = \left(-\frac{1}{e^3} \right) \exp(1/e) < 0.$$

Thus f has a maximum at $x = e$. For $x = 0$ we find that f takes the form $f(x = 0) = 0$ (L'Hospital's rule). Furthermore we have

$$\lim_{x \to \infty} \sqrt[x]{x} = 1.$$

Problem 10.   Let $a_1, a_2, \ldots, a_n$ be positive numbers and $b_1, b_2, \ldots, b_n$ be nonnegative numbers such that

$$\sum_{j=1}^{n} b_j > 0.$$

Show that (*log-sum inequality*)

$$\sum_{j=1}^{n} \left(a_j \ln \frac{a_j}{b_j} \right) \geq \left(\sum_{j=1}^{n} a_j \right) \ln \frac{\left(\sum_{j=1}^{n} a_j \right)}{\left(\sum_{j=1}^{n} b_j \right)}$$

with the conventions based on continuity arguments

$$0 \cdot \ln 0 = 0, \qquad 0 \cdot \ln \frac{p}{0} = \infty, \quad p > 0.$$

Show that equality holds if and only if $a_j/b_j = $ constant for all $j = 1, 2, \ldots, n$.

Solution 10.   We set

$$A := \sum_{j=1}^{n} a_j, \qquad B := \sum_{j-1}^{n} b_j.$$

Then

$$\sum_{j=1}^{n}\left(a_j \ln \frac{a_j}{b_j}\right) - \left(\sum_{j=1}^{n} a_j\right) \ln \frac{\sum_{j=1}^{n} a_j}{\sum_{j=1}^{n} b_j} = \sum_{j=1}^{n}\left(a_j \ln \frac{a_j}{b_j}\right) - A \ln \frac{A}{B}$$

$$= A\sum_{j=1}^{n} \frac{a_j}{A} \ln \frac{a_j/A}{b_j/B}$$

$$\geq A\sum_{j=1}^{n} \left(\frac{a_j}{A}\left(1 - \frac{b_j/B}{a_j/A}\right)\right)$$

$$= A\sum_{j=1}^{n} \left(\frac{a_j}{A} - \frac{b_j}{B}\right)$$

$$= A(1-1) = 0.$$

The inequality comes from $1 - 1/x \leq \ln x$ $(x > 0)$ with equality iff $x = 1$. For equality to hold, we need to have

$$\frac{a_j/A}{b_j/B} = 1 \quad \text{for all } j = 1, 2, \ldots, n.$$

This implies

$$\frac{a_j}{b_j} = \text{constant} \quad \text{for all } j = 1, 2, \ldots, n.$$

Problem 11.   Let n be a positive integer and $a, b \geq c/2 > 0$. Show that

$$|a^{-n} - b^{-n}| \leq 4nc^{-n-1}|a - b|.$$

Solution 11.   We have

$$|a^{-n} - b^{-n}| = \left|\frac{b^n - a^n}{(ab)^n}\right|$$

$$= \frac{\sum_{i=1}^{n} b^{n-i}a^{i-1}}{(ab)^n}|b - a|$$

$$= |a - b|\sum_{i=1}^{n} b^{-i}a^{i-n-1}$$

$$\leq 4nc^{-n-1}|b - a|.$$

Chapter 19

Functional Analysis

Problem 1. Let F be a closed set on a complete metric space X. A *contracting mapping* is a mapping $f : F \to F$ such that

$$d(f(x), f(y)) \le kd(x, y), \qquad 0 \le k < 1, \quad d \text{ distance in } X .$$

One also says that f is *Lipschitzian* of order $k < 1$.

(i) Prove the following theorem: A contracting mapping f has strictly one fixed point, i.e. there is one and only one point x^* such that $x^* = f(x^*)$.

(ii) Apply the theorem to the linear equation $A\mathbf{x} = \mathbf{b}$, where A is an $n \times n$ matrix over the real numbers and $\mathbf{x} = (x_1, x_2, \dots, x_n)^T$.

Solution 1. (i) The proof is by iteration. Let $x_0 \in F$. Then

$$f(x_0) \in F, \dots, f^{(n)}(x_0) = f(f^{(n-1)}(x_0)) \in F$$

and

$$d(f^{(n)}(x_0), f^{(n-1)}(x_0)) \le kd(f^{(n-1)}(x_0), f^{(n-2)}(x_0))$$
$$\le \cdots \le k^{n-1}d(f(x_0), x_0).$$

Since $k < 1$ the sequence $f^{(n)}$ is a Cauchy sequence and tends to a limit x^* when n tends to infinity

$$x^* = \lim_{n\to\infty} f^{(n)}(x_0) = \lim_{n\to\infty} f(f^{(n-1)}(x_0)) = f(x^*).$$

The uniqueness of x^* results from the defining property of contracting mappings: Assume that there is another point y^* such that $y^* = f(y^*)$, then on the one hand

$$d(f(y^*), f(x^*)) = d(y^*, x^*)$$

but on the other hand $d(f(y^*), f(x^*)) \leq kd(y^*, x^*)$, $k < 1$. Therefore $d(y^*, x^*) = 0$ and $y^* = x^*$.

(ii) We set $c_{jk} := -a_{jk} + \delta_{jk}$, where a_{jk} are the matrix elements of A and δ_{jk} is the Kronecker delta. Then the linear equation $A\mathbf{x} = \mathbf{b}$ takes the form $\mathbf{x} = C\mathbf{x} + \mathbf{b}$. If

$$\sum_{k=1}^{n} |c_{jk}| < 1, \qquad j = 1, 2, \ldots, n$$

then the theorem can be applied.

Problem 2.   Let A and B be two arbitrary $n \times n$ matrices over $\mathbb{R}$. Define

$$(A, B) := \operatorname{tr}(AB^T) \tag{1}$$

where B^T denotes the transpose of B.
(i) Show that

$$(A, A) \geq 0 \tag{2a}$$

$$(A, B) = (B, A) \tag{2b}$$

$$(cA, B) = c(A, B) \tag{2c}$$

$$(A_1 + A_2, B) = (A_1, B) + (A_2, B) \tag{2d}$$

where $c \in \mathbb{R}$.
(ii) Let O be an orthogonal $n \times n$ matrix over $\mathbb{R}$, i.e. $O^{-1} = O^T$. Find (O, O).
(iii) Can the result be extended to infinite-dimensional matrices?

Solution 2.   (i) We find

$$(A, A) = \operatorname{tr}(AA^T) = \sum_{j=1}^{n}\sum_{k=1}^{n} a_{jk}a_{jk}.$$

Since $a_{jk}a_{jk} \geq 0$ we obtain inequality (2a). Owing to

$$(A, B) = \operatorname{tr}(AB^T) = \sum_{j=1}^{n}\sum_{k=1}^{n} a_{jk}b_{jk}$$

and

$$(B, A) = \operatorname{tr}(BA^T) = \sum_{j=1}^{n}\sum_{k=1}^{n} b_{jk}a_{jk}$$

we obtain (2b). Since

$$(cA, B) = \operatorname{tr}(cAB^T) = c\operatorname{tr}(AB^T) = c(A, B)$$

we find (2c). To prove (2d) we recall that $\operatorname{tr}(X+Y) = \operatorname{tr}(X) + \operatorname{tr}(Y)$ for any $n \times n$ matrices X and Y. Therefore

$$(A_1+A_2, B) = \operatorname{tr}((A_1+A_2)B^T) = \operatorname{tr}(A_1 B^T)+\operatorname{tr}(A_2 B^T) = (A_1, B)+(A_2, B).$$

Consequently, definition (1) defines a scalar product for the vector space of $n \times n$ matrices over the field of real numbers.

(ii) Since $O^T = O^{-1}$ for an orthogonal matrix we find

$$(O, O) = \operatorname{tr}(OO^T) = \operatorname{tr}(OO^{-1}) = \operatorname{tr}(I_n) = n.$$

(iii) The results can be extended when we impose the condition

$$\sum_{j=1}^{\infty}\sum_{k=1}^{\infty} |a_{jk}|^2 < \infty.$$

If an infinite-dimensional matrix A satisfies condition (3) we call A a *Hilbert-Schmidt operator*. The norm of a Hilbert-Schmidt operator A is given by

$$\|A\| := \sqrt{\sum_{j=1}^{\infty}\sum_{k=1}^{\infty} |a_{jk}|^2}.$$

Problem 3. Show that if $a < b$, and $n \in \mathbb{Z}$ the functions

$$f_n(x) = e^{inx} \tag{1}$$

are a linearly independent set on the interval $[a, b]$.

Solution 3. If $b - a \geq 2\pi$ and

$$\sum_{n=p}^{q} c_n e^{inx} = 0$$

on $[a, b]$, then the orthogonality relationships among the functions f_n, i.e.

$$\int_0^{2\pi} e^{i(n-m)x}dx = 2\pi\delta_{mn}$$

imply that all c_n are equal to zero. Here δ_{mn} denotes the Kronecker delta. If $0 < b - a < 2\pi$ and

$$g(x) = \sum_{n=p}^{q} c_n e^{inx} = 0$$

on $[a, b]$, then g has an extension to an entire function on $\mathbb{C}$. Thus

$$g(x) = 0, \qquad \sum_{n=p}^{q} c_n e^{inx} = 0$$

on $[0, 2\pi]$ and the previous argument shows all c_n are equal to zero.

Problem 4.   (i) Let $I = [0, 1]$ with Lebesgue measure and $x \in I$. Let the nth *Rademacher function* r_n be defined as

$$r_n(x) := \mathrm{sgn}(\sin(2^n \pi x))$$

where $n = 0, 1, 2, \ldots$ and sgn denotes the signum function. Show that the Rademacher functions r_n constitute a linearly independent set.
(ii) For $m \in \mathbb{N}$ there is in $\mathbb{N} \cup \{0\}$ a unique finite set $\{n_1, n_2, \ldots, n_{K_m}\}$ such that

$$m = \sum_{k=1}^{K_m} 2^{n_k}.$$

The m-th *Walsh function* is defined by

$$W_m(x) := \prod_{k=1}^{K_m} r_{n_k}(x)$$

where $x \in I \equiv [0, 1]$. Show that $\{W_m\}_{m=1}^{\infty}$ is a complete orthogonal set in the Hilbert space $L_2(I)$. This is the Hilbert space of the square-integrable functions.

Solution 4.   (i) The function r_n partitions $[0, 1]$ into three sets, $S_n^{\pm}$ and S_n^0, where

$$S_n^{\pm} = r_n^{-1}(\pm 1) \qquad S_n^0 = r_n^{-1}(0).$$

Furthermore,

$$S_0^+ = (0, 1) \qquad S_0^- = \emptyset \qquad S_0^0 = \{0, 1\}.$$

If $n \geq 1$ each of $S_n^{\pm}$ consists of 2^{n-1} disjoint intervals, each of length 2^{-n} and S_n^0 consists of their $2^n + 1$ endpoints. The intervals of S_n^+ alternate with those of S_n^-. If $m > n$ the intervals of $S_m^{\pm}$ equipartition those of $S_n^{\pm}$ from which the linear independence of the Rademacher functions r_n, $n = 1, 2, \ldots$ follows. Obviously, $\{r_n\}_n$ is an orthonormal set in the Hilbert space $L_2(I)$. However, the Rademacher functions do not form a basis in the Hilbert space $L_2(I)$.

(ii) The result of (i) shows that

$$\{W_m\}_{m=1}^{\infty}$$

is an orthonormal set in the Hilbert space $L_2(I)$. Let

$$f \in (\{W_m\}_{m=1}^{\infty})^{\perp}$$

where $M^{\perp}$ is the *orthogonal complement* of M. The set $M^{\perp}$ denotes all vectors which are orthogonal to every vector in M. Then for all $N \in \mathbb{N}$ and all x,

$$F(x) = \int_0^1 f(t) \prod_{m=0}^{N} (1 + r_m(x)r_m(t))dt = 0.$$

Induction shows that if $N > 1$ and

$$\frac{k}{2^m} \le x \le \frac{(k+1)}{2^m}$$

then

$$\prod_{m=0}^{N} (1 + r_m(x)r_m(t)) \neq 0$$

if and only if

$$\frac{k}{2^N} \le t \le \frac{(k+1)}{2^N}.$$

Obviously $1 + r_2(x)r_2(t) \neq 0$ if and only if x and t are in the same half of I,

$$(1 + r_2(x)r_2(t))(1 + r_3(x)r_3(t)) \neq 0$$

if and only if x and t are in the same quarter of I, etc. Hence

$$\int_{k/2^N}^{(k+1)/2^N} f(t)dt = 0$$

for all k, $N \in \mathbb{N}$. Consequently, $f(t) = 0$ almost everywhere. It follows that

$$\{W_m\}_{m=1}^{\infty}$$

is a complete orthornormal system. In other words the Walsh functions form an orthonormal basis in the Hilbert space $L_2(I)$.

Problem 5. Let P and Q be two linear operators in a normed space. Let a be a nonzero real or complex number and let I be the identity operator. Show that if all iterated operators Q^m exist then the relation

$$PQ - QP = aI \qquad (1)$$

cannot be satisfied by two bounded operators P, Q in a normed space.

Solution 5.   We assume that the iterated operators Q^m $(m = 0, 1, 2 \ldots)$ are meaningful. From (1) it follows that P and Q can neither be the null operator nor a constant. Consequently,

$$\|P\| \neq 0, \qquad \|Q\| \neq 0$$

where $\|.\|$ denotes the norm. From (1) we find by induction that

$$PQ^n - Q^n P = anQ^{n-1}. \tag{2}$$

For $n = 1$, (2) is obviously true. It follows then that

$$\begin{aligned} PQ^{n+1} - Q^{n+1}P &= (PQ^n - Q^n P)Q + Q^n(PQ - QP) \\ &= anQ^{n-1}Q + Q^n aI \\ &= a(n+1)Q^n \end{aligned}$$

which completes the induction. We assume now that P and Q are bounded operators with the norms $\|P\|$ and $\|Q\|$. Taking norms on both sides of this equation for $n = 1, 2, \ldots$ we find

$$|a|n\|Q^{n-1}\| = \|PQ^n - Q^n P\| \leq \|P\|\,\|Q^n\| + \|Q^n\|\,\|P\| = 2\|P\|\,\|Q^n\|.$$

Thus

$$|a|n\|Q^{n-1}\| \leq 2\|P\|\,\|Q^{n-1}\|\,\|Q\|.$$

Let N be an integer with

$$N > \frac{2}{|a|}\|P\|\,\|Q\|.$$

We have to study two cases:

Case 1. Let $\|Q^{N-1}\| \neq 0$. This result is in contradiction for $n = N$.

Case 2: Let $\|Q^{N-1}\| = 0$. Owing to

$$|a|\,n\,\|Q^{n-1}\| \leq 2\|P\|\,\|Q^n\|$$

for all $n = 1, 2, 3, \ldots$ we obtain

$$\|Q^{N-2}\| = 0, \quad \|Q^{N-3}\| = 0, \quad \ldots, \quad \|Q^1\| = 0.$$

Again we find a contradiction.

Owing to this result there are no two finite dimensional matrices P and Q which satisfy (1) with $a \neq 0$. This also follows from $\operatorname{tr}([A, B]) = 0$ for $n \times n$ matrices A and B over the field of the complex numbers. Here $[\, ,\,]$ denotes the commutator. There are *unbounded operators* which satisfy (1). Consider the infinite-dimensional matrix

$$b := \begin{pmatrix} 0 & \sqrt{1} & 0 & 0 & \\ 0 & 0 & \sqrt{2} & 0 & \\ 0 & 0 & 0 & \sqrt{3} & \\ & \ddots & & & \ddots \\ & & & \ddots & \\ & & & & \ddots \end{pmatrix} .$$

Let b^T be the transpose of b. Then $[b, b^T] = I$, where I is the infinite-dimensional unit matrix and $[\, ,\,]$ denotes the commutator. The operator $b^T b$ is given by the infinite-dimensional diagonal matrix (unbounded operator)

$$b^T b = \operatorname{diag}(0, 1, 2, \ldots) .$$

Problem 6. Let f_1, f_2, $\ldots$, f_n be continuous real-valued functions on the interval $[a, b]$. Show that the set $\{\, f_1,\, f_2, \ldots, f_n \,\}$ is linearly dependent on $[a, b]$ if and only if

$$\det \left(\int_a^b f_i(x) f_j(x) dx \right) = 0 .$$

Solution 6. Let A be the matrix with the entries

$$A_{ij} := \int_a^b f_i(x) f_j(x) dx .$$

If the determinant of the matrix A vanishes, the matrix A is singular. Let $\mathbf{a}$ be a nonzero (column) n-vector with $A\mathbf{a} = \mathbf{0}$. Then

$$0 = \mathbf{a}^T A \mathbf{a} = \sum_{i=1}^n \sum_{j=1}^n \int_a^b a_i f_i(x) a_j f_j(x) dx = \int_a^b \left(\sum_{i=1}^n a_i f_i(x) \right)^2 dx .$$

Since the f_i's are continuous functions, the linear combinations

$$\sum_{j=1}^n a_j f_j$$

must vanish identically. Hence, the set $\{\, f_i \,\}$ is linearly dependent on $[a, b]$. Conversely, if $\{\, f_i \,\}$ is linearly dependent, some f_i can be expressed as a linear combination of the rest, so some row of A is a linear combination of the rest and A is singular.

Problem 7.   Consider the inner product space

$$C[a, b] = \{\, f(x) \,:\, f \text{ is continuous on } x \in [a, b] \,\}$$

with the inner product

$$\langle f, g \rangle := \int_a^b f(x)g^*(x)dx \,.$$

This implies a norm

$$\langle f, f \rangle = \int_a^b f(x)f^*(x)dx = \|f\|^2 \,.$$

Show that $C[a, b]$ is incomplete. This means find a *Cauchy sequence* in the space $C[a, b]$ which converges to an element which is not in the space $C[a, b]$.

Solution 7.   Consider the sequence of continuous functions

$$g_k(x) = \frac{1}{2} + \frac{1}{\pi} \arctan(kx), \qquad -1 \le x \le 1$$

i.e. $a = -1$ and $b = 1$. We find

$$\lim_{k,p\to\infty} \|g_k - g_p\|^2 = \lim_{k,p\to\infty} \int_{-1}^{+1} (g_k(x) - g_p(x))^2 dx = 0 \,.$$

In other words, the sequence is a Cauchy sequence. For fixed x we have

$$\lim_{k\to\infty} g_k(x) = g(x) = \begin{cases} 1 & 0 < x \le 1 \\ 1/2 & x = 0 \\ 0 & -1 \le x < 0 \end{cases}$$

which is a discontinuous function, i.e., $g \notin C[-1, 1]$. We say that the sequence of functions $\{\, g_k \,:\, k = 1, 2, \ldots \,\}$ converges pointwise to the function g. We also have

$$\lim_{k\to\infty} \|g - g_k\|^2 = \lim_{k\to\infty} \int_{-1}^{+1} (g(x) - g_k(x))^2 dx = 0 \,.$$

Thus the limit of the Cauchy sequence does not lie in the inner product space $C[-1, 1]$. Thus $C[a, b]$ is incomplete.

Problem 8. Consider the Hilbert space $L_2[0,1]$. Find a non-trivial polynomial p

$$p(x) = ax^3 + bx^2 + cx + d$$

such that $\langle p, 1 \rangle = 0$, $\langle p, x \rangle = 0$, $\langle p, x^2 \rangle = 0$.

Solution 8. Using the ansatz for a polynomial p we obtain

$$\langle p, 1 \rangle = \int_0^1 p(x)dx = \frac{a}{4} + \frac{b}{3} + \frac{c}{2} + d = 0$$

$$\langle p, x \rangle = \int_0^1 p(x)x\,dx = \frac{a}{5} + \frac{b}{4} + \frac{c}{3} + \frac{d}{2} = 0$$

$$\langle p, x^2 \rangle = \int_0^1 p(x)x^2\,dx = \frac{a}{6} + \frac{b}{5} + \frac{c}{4} + \frac{d}{3} = 0.$$

This system can be written in matrix form

$$\begin{pmatrix} 1/4 & 1/3 & 1/2 \\ 1/5 & 1/4 & 1/3 \\ 1/6 & 1/5 & 1/4 \end{pmatrix} \begin{pmatrix} a \\ b \\ c \end{pmatrix} = \begin{pmatrix} -d \\ -d/2 \\ -d/3 \end{pmatrix}.$$

The inverse of the matrix on the left-hand side exists. If we assume that $d \neq 0$ we find a non-trivial polynomial.

Problem 9. Let $C^m[a,b]$ be the vector space of real-valued m-times differentiable functions and the m-th derivative is continuous over the interval $[a,b]$ $(b > a)$. We define an inner product (scalar product) of two such functions f and g as

$$\langle f, g \rangle_m := \int_a^b \left(fg + \frac{df}{dx}\frac{dg}{dx} + \cdots + \frac{d^m f}{dx^m}\frac{d^m g}{dx^m} \right) dx.$$

Given (Legendre polynomials)

$$f(x) = \frac{1}{2}(3x^2 - 1), \qquad g(x) = \frac{1}{2}(5x^3 - 3x)$$

and the interval $[-1,1]$, i.e. $a = -1$ and $b = 1$. Show that f and g are orthogonal with respect to the inner product $\langle f, g \rangle_0$. Are they orthogonal with respect to $\langle f, g \rangle_1$?

Solution 9. Straightforward calculation yields

$$\langle f, g \rangle_0 = \frac{1}{4}\int_{-1}^1 (3x^2 - 1)(5x^3 - 3x)dx = \frac{1}{4}\int_{-1}^1 (15x^5 - 14x^3 + 3x)dx = 0.$$

We have $df/dx = 3x$, $dg/dx = 15x^2/2 - 3/2$. Then since $\langle f, g \rangle_0 = 0$ we have

$$\langle f, g \rangle_1 = \int_{-1}^{1} 3x \left(\frac{15}{2}x^2 - \frac{3}{2} \right) dx = \int_{-1}^{1} \left(\frac{45}{2}x^3 - \frac{9}{2}x \right) dx = 0 \, .$$

Problem 10. Let $L_2[0, 1]$ be the Hilbert space of the square-integrable functions in the Lebesgue sense over the unit interval $[0, 1]$. Let $f : [0, 1] \mapsto [0, 1]$ be

$$f(x) = \begin{cases} 2x & \text{for} \quad 0 \leq x \leq 1/2 \\ 2(1 - x) & \text{for} \quad 1/2 < x \leq 1. \end{cases} \tag{1}$$

A basis in $L_2[0, 1]$ is given by

$$\{ u_n(x) := e^{2\pi i x n} \; : \; n \in \mathbb{Z} \} \, .$$

Obviously, $f \in L_2[0, 1]$. Then f can be expanded in a Fourier series with respect to u_n. Find the Fourier coefficients.

Solution 10. The *Fourier expansion* of f is given by

$$f(x) = \sum_{n \in \mathbb{Z}} \langle f, u_n \rangle u_n \equiv \sum_{n \in \mathbb{Z}} c_n u_n$$

where $(,)$ denotes the scalar product in the Hilbert space $L_2[0, 1]$. The scalar product is defined by

$$\langle f, g \rangle := \int_0^1 f(x) \bar{g}(x) dx$$

where $\bar{g}$ is the complex conjugate of g. Hence the expansion coefficients c_n are given by

$$c_n \equiv \langle f, u_n \rangle := \int_0^1 f(x) \bar{u}_n(x) dx = \int_0^1 f(x) e^{-2\pi i x n} dx$$

or

$$c_n = 2 \int_0^{1/2} x e^{-2\pi i x n} dx + 2 \int_{1/2}^1 e^{-2\pi i x n} dx - 2 \int_{1/2}^1 x e^{-2\pi i x n} dx.$$

For $n = 0$ we find $c_0 = 1/2$. For $a \neq 0$ we have

$$\int e^{ax} dx = \frac{1}{a} e^{ax}, \qquad \int x e^{ax} dx = \frac{1}{a} e^{ax} x - \frac{1}{a^2} e^{ax}.$$

Using these two integrals we find for $n \neq 0$ that $c_n = 4(1 - (-1)^n)/a^2$, where $a = -2\pi i n$.

Chapter 20

Combinatorics

In combinatorics we study the enumeration, combination, and permutations of sets of elements.

Problem 1. (i) In how many ways f_n can a group of n persons arrange themselves in a row of n chairs?

(ii) In how many ways f_n can a group of n persons arrange themselves around a circular table.

Solution 1. (i) The n persons can arrange themselves in a row in

$$n(n-1)(n-2)\cdots 2 \cdot 1 = n! \tag{1}$$

ways. In other words: one person can sit in n different chairs. Therefore

$$f_n = nf_{n-1} \tag{2}$$

with $n = 2, 3, \ldots$ and the initial condition is given by $f_1 = 1$. The solution of the difference equation (2) with the initial condition is given by (1).

(ii) One person can sit in any place at the circular table. The other $n-1$ persons can then arrange themselves in

$$f_n = (n-1)(n-2)\cdots 2 \cdot 1 = (n-1)!$$

In other words n objects can be arranged in a circle in $(n-1)!$ ways.

Problem 2. What is the number A_n of ways of going up n steps, if we may take one or two steps at a time? Determine

$$\sum_{n=0}^{\infty} A_n x^n.$$

Solution 2.   We may begin by taking one or two steps. In the first case, we have A_{n-1} possibilities to continue; in the second, we have A_{n-2} possibilities. Thus

$$A_n = A_{n-1} + A_{n-2} \tag{1}$$

where $A_1 = 1$ and $A_2 = 2$. The sequence A_n is the sequence of the Fibonacci numbers. We set $A_0 = 1$ and

$$f(x) = \sum_{n=0}^{\infty} A_n x^n$$

as the generating function. Then

$$xf(x) = \sum_{n=1}^{\infty} A_{n-1} x^n, \qquad x^2 f(x) = \sum_{n=2}^{\infty} A_{n-2} x^n$$

and using (1) we find

$$f(x) - xf(x) - x^2 f(x) = A_0 + (A_1 - A_0)x + \sum_{n=2}^{\infty} (A_n - A_{n-1} - A_{n-2})x^n.$$

Thus

$$f(x) - xf(x) - x^2 f(x) = A_0 = 1.$$

Consequently, the generating function is given by

$$f(x) = \frac{1}{1 - x - x^2}.$$

To obtain an explicit formula for A_n we write this as

$$f(x) = \frac{1}{\sqrt{5}x} \left(\frac{1}{1 - \frac{1+\sqrt{5}}{2}x} - \frac{1}{1 - \frac{1-\sqrt{5}}{2}x} \right)$$

$$= \frac{1}{\sqrt{5}x} \sum_{k=0}^{\infty} \left(\left(\frac{1+\sqrt{5}}{2}x \right)^k - \left(\frac{1-\sqrt{5}}{2}x \right)^k \right)$$

$$2pt] = \frac{1}{\sqrt{5}} \sum_{n=0}^{\infty} \left(\left(\frac{1+\sqrt{5}}{2} \right)^{n+1} - \left(\frac{1-\sqrt{5}}{2} \right)^{n+1} \right) x^n.$$

Therefore

$$A_n = \frac{1}{\sqrt{5}}\left(\left(\frac{1+\sqrt{5}}{2}\right)^{n+1} - \left(\frac{1-\sqrt{5}}{2}\right)^{n+1}\right)$$

where $n = 0, 1, 2, \ldots$.

Problem 3. We have n Rand. Every day we buy exactly one of the following products: pretzel (1 Rand), candy (2 Rand), ice-cream (2 Rand). What is the number B_n of possible ways of spending all the money?

Solution 3. On the first day, we have three choices: we may buy a pretzel, in which case we have B_{n-1} further possible ways to spend the remaining $n-1$ Rand; or we may buy candy for 2 Rand, and then we can spend the rest in B_{n-2} ways; similarly we have B_{n-2} possibilities if we first buy an ice-cream. Thus we find the linear difference equation with constant coefficients

$$B_n = B_{n-1} + 2B_{n-2} \qquad (n \geq 3). \tag{1}$$

The difference equation (1) is of second order. Thus we need two initial conditions. Obviously,

$$B_1 = 1, \qquad B_2 = 3 \tag{2}$$

are the initial values of the linear difference equation (1). The first few terms are

$$B_3 = 5, \quad B_4 = 11, \quad B_5 = 21, \quad B_6 = 43.$$

Since (1) is a linear difference equation with constant coefficients we can solve it with the ansatz

$$B_n = ar^n.$$

We obtain the quadratic equation $r^2 = r + 2$. Consequently, the solution of (1) with the initial conditions (2) is given by

$$B_n = \frac{1}{3}(2^{n+1} + (-1)^n).$$

Problem 4. Let $A_1, A_2, \ldots, A_p$ be finite sets. Let $|A_k|$ be the number of elements of A_k. Show that

$$\left|\bigcup_{j=1}^{p} A_j\right| = \sum_{j=1}^{p} |A_j| - \sum_{1 \leq j < k \leq p} |A_j \cap A_k| + \cdots + (-1)^{p+1}\left|\bigcap_{j=1}^{p} A_j\right|. \tag{1}$$

Solution 4. The proof applies induction on $p \geq 2$. For $p = 2$ we find that (1) takes the form

$$|A_1 \cup A_2| = |A_1| + |A_2| - |A_1 \cap A_2|$$

which is obviously true. Suppose the equation is true for each union of at most $p - 1$ sets. It follows that

$$|A_1 \cup A_2 \cdots \cup A_p| = |A_1 \cup A_2 \cdots \cup A_{p-1}| + |A_p| - |(A_1 \cup A_2 \cdots \cup A_{p-1}) \cap A_p|.$$

Applying the *distributive law for intersections* of sets we find

$$(A_1 \cup A_2 \cdots \cup A_{p-1}) \cap A_p = (A_1 \cap A_p) \cup (A_2 \cap A_p) \cup \cdots \cup (A_{p-1} \cap A_p)$$

and from the inductive hypothesis it follows that

$$|A_1 \cup A_2 \cdots \cup A_p| = \sum_{1 \leq j < p} |A_j| - \sum_{1 \leq j < k < p} |A_j \cap A_k| + \cdots (-1)^p \left| \bigcap_{j=1}^{p-1} A_j \right|$$

$$+ |A_p| - \sum_{1 \leq j < p} |A_j \cap A_p| + \cdots (-1)^{p+1} \left| \bigcap_{j=1}^{p} A_j \right|$$

where we have used the *idempotent law of intersection* in the form

$$(A_j \cap A_p) \cap (A_k \cap A_p) = A_j \cap A_k \cap A_p, \quad \cdots \quad \bigcap_{j=1}^{p-1} (A_j \cap A_p) = \bigcap_{j=1}^{p} A_j.$$

By regrouping terms, we obtain (1). Equation (1) is called the *principle of inclusion and exclusion*.

Problem 5. Show that the number of arrangements N of a set of n objects in p boxes such that the jth box contains n_j objects, for $j = 1, \ldots, p$ is equal to

$$N = \frac{n!}{n_1! n_2! \cdots n_p!}$$

where $n_j \geq 0$ and

$$n_1 + n_2 + \cdots + n_p = n.$$

Solution 5. The n_1 objects in the first box can be chosen in

$$\binom{n}{n_1}$$

ways, the n_2 objects in the second box can be chosen from the $n - n_1$ remaining objects in

$$\binom{n - n_1}{n_2}$$

ways, etc. The total number of arrangements is equal to

$$\binom{n}{n_1}\binom{n - n_1}{n_2}\binom{n - n_1 - n_2}{n_3}\cdots\binom{n - n_1 - \cdots - n_{p-1}}{n_p} \equiv \frac{n!}{n_1! n_2! \ldots n_p!}.$$

Problem 6. A derangement (or fixed-point-free permutation) of $\{1, 2, \ldots, n\}$ is a permutation f such that $f(j) \neq j$ for all $j = 1, 2, \ldots, n$. What is the number of derangements of n objects?

Solution 6. There are $n!$ permutations. Using

$$\binom{n}{n} = 1, \qquad \binom{n}{1} = (n - 1)!$$

we find that the number of derangements of n objects is given by

$$n! - n(n - 1)! + \binom{n}{2}(n - 2)! - + \cdots + (-1)^n = n! \sum_{j=2}^{n}(-1)^j \frac{1}{j!}.$$

Problem 7. *Bell numbers* are the sequence

$$\{1, 1, 2, 5, 15, 52, 203, 877, 4140, \ldots\}.$$

The numbers count (starting from 0) the ways that n distinguishable objects can be grouped into sets if not set can be empty. For example the letters $A\ B\ C$ can be grouped into sets so that

(1) A, B, C are in three separate sets: $\{A\}, \{B\}, \{C\}$
(2) A and B are together and C is separate: $\{A, B\}, \{C\}$
(3) A and C together and B separate: $\{A, C\}, \{B\}$
(4) B and C together and A separate: $\{B, C\}, \{A\}$
(5) A, B, C are all together in a single set: $\{A, B, C\}$

Thus for $n = 3$ there are five partitions. Thus the third Bell number is 5.
(i) Let P_n denote the nth Bell number, i.e. the number of all partitions of n objects. Then we have

$$P_n = \frac{1}{e} \sum_{k=0}^{\infty} \frac{k^n}{k!}.$$

Find a recurrence relation for P_n.

(ii) Find the MacLaurin expansion (expansion around 0) of $\exp(\exp(x))$ and establish a connection with the Bell numbers.

Solution 7.   (i) Let S be the set to be partioned and $x \in S$. If the class containing x has k elements, it can be chosen in

$$\binom{n-1}{k-1}$$

ways and the remaining $n-k$ elements can be partioned in P_{n-k} ways. Thus the number of partitions in which the class containing x has k elements is

$$\binom{n-1}{k-1} P_{n-k} .$$

This remains true for $k = n$ if we set $P_0 = 1$. Thus

$$P_n = \sum_{k=1}^{n} \binom{n-1}{k-1} P_{n-k} = \sum_{k=0}^{n-1} \binom{n-1}{k} P_k .$$

(ii) Since $d(\exp(\exp(x)))/dx = \exp(x)\exp(\exp(x))$ and $\exp(0) = 1$ we have

$$\exp(\exp(x)) = \sum_{j=0}^{\infty} \frac{e^{jx}}{j!}$$

$$= \sum_{k=0}^{\infty}\sum_{j=0}^{\infty} \frac{(jx)^k}{j!k!}$$

$$= \sum_{k=0}^{\infty} \left(\sum_{j=0}^{\infty} \frac{j^k}{j!} \right) \frac{x^k}{k!}$$

$$= \exp(\sum_{k=0}^{\infty} P_k \frac{x^k}{k!} .$$

Thus we find

$$\exp(\exp(x)) = e \left(1 + \frac{1x}{1!} + \frac{2x^2}{2!} + \frac{5x^3}{3!} + \cdots \right) .$$

Thus the coefficients 1, 1, 2, 5, ... are the Bell numbers.

Problem 8.   Consider the n-dimensional unit cube in $\mathbb{R}^n$. How many k-dimensional surfaces ($1 \le k < n$) does the n-dimensional unit cube have?

Solution 8. We have

$$\binom{n}{k} 2^{n-k}.$$

For example for $n = 3$ and $k = 2$ we have 6 surfaces. For $n = 3$ and $k = 1$ we have 12 (lines).

Problem 9. How many arrangements of $a, a, a, b, b, b, c, c, c$ are there so that no 2 letters of the same type appear consecutively? For example, *abcabcabc* and *cabcabcab* would be such arrangements.

Solution 9. The total number of arrangements of $a, a, a, b, b, b, c, c, c$ is

$$\frac{9!}{3!3!3!} = 1680.$$

This follows from the formula for the permutations of multisets. We use the inclusion-exclusion principle to find the number of arrangements. Let S_a be the multiset of those arrangements that have two a's consecutive. Similarly, we define S_b and S_c. If two a's are consecutive, then glue them to form a block which we consider as one letter. Then $|S_a|$ is the number of ways to permute the multiset

$$\{aa, b, b, b, c, c, c\}$$

which is $7!/(3!3!)$, times the number of distinct ways to insert the last a, namely 7. When inserting the last a we only count the space to the left of aa and not to the right as this would be an over count. Thus

$$|S_a| = 7 \cdot \frac{7!}{3!3!} = 980.$$

Similarly, $|S_b| = |S_c| = S_a|$ by symmetry. Next we construct $S_a \cap S_b$ by finding all multiset permutations on $\{aa, bb, c, c, c\}$ and then inserting an a and a b. There are 5 distinct ways to insert the extra a into the 5-perm, 6 distinct ways to insert the extra b into the 6-perm. In addition, it is possible that a is inserted just to the right of aa and then it is separated by the b on the second step. This is equivalent in finding a multiset permutation of $\{aaba, bb, c, c, c\}$. Therefore

$$|S_a \cap S_b| = \binom{5}{1,1,3} \cdot 5 \cdot 6 + \binom{5}{1,1,3} = 620.$$

Analogously, we find

$$|S_c \cap S_b| = |S_a \cap S_c| = 620.$$

Finally, we calculate the number of elments in the multiset $S_a \cap S_b \cap S_c$, i.e. the number of distinct multiset permutations of $\{aa, a, bb, b, cc, c\}$. There are 6 permutations of $\{aa, bb, cc\}$. We can add the extra a in 4 ways, either to the right of aa or in one of the 3 other positions. In the first case, after inserting the b and c we must end up with one of the strings $aaba$ (with 3 places to put the c) $aaca$ (with three places to put the b) $aabca$ or $aacba$, totally 8 possibilities. In the second case, if the b goes to the right of the bb the c is forced, if the b goes in any of the remaining 4 positions, the c can be inserted anywhere except to the right of cc, giving $3 + 3 \cdot 4 \cdot 5 = 63$ possibilities. Therefore

$$|S_a \cap S_b \cap S_c| = 6 \cdot (8 + 63) = 426 \,.$$

Thus using the inclusion-exclusion principle the number of mutiset permutations with no double letters is

$$1680 - 3 \cdot 980 + 3 \cdot 620 - 426 = 174 \,.$$

Chapter 21

Convex Sets and Functions

A subset C of $\mathbb{R}^n$ is said to be *convex* if for any $\mathbf{a}$ and $\mathbf{b}$ in C and any θ in $\mathbb{R}$, $0 \le \theta \le 1$, the n-tuple $\theta \mathbf{a} + (1 - \theta)\mathbf{b}$ also belongs to C. In other words, if $\mathbf{a}$ and $\mathbf{b}$ are in C, then

$$\{\theta \mathbf{a} + (1 - \theta)\mathbf{b} \,:\, 0 \le \theta \le 1\} \subset C.$$

Let C be a convex subset of $\mathbb{R}^n$. Let f be a real-valued function with domain C. If for each $\mathbf{x}$ and $\mathbf{y}$ in C and each θ, $0 \le \theta \le 1$, the inequality

$$f(\theta \mathbf{x} + (1 - \theta)\mathbf{y}) \le \theta f(\mathbf{x}) + (1 - \theta)f(\mathbf{y})$$

holds, then f is said to be a *convex function*. If $-f$ is convex, then f is called a *concave function*.

Problem 1. Let B^n be the unit ball in $\mathbb{R}^n$, i.e.

$$B^n := \{(a_1, a_2, \ldots, a_n) \in \mathbb{R}^n \,:\, \sum_{j=1}^{n} a_j^2 \le 1\}.$$

Show that B^n is convex.

Solution 1. Let $\mathbf{a} \in B^n$, $\mathbf{b} \in B^n$ and

$$c_j := \theta a_j + (1 - \theta)b_j$$

229

where $j = 1, 2, \ldots, n$. Then we have

$$\sum_{j=1}^{n} c_j^2 = \sum_{j=1}^{n} (\theta a_j + (1-\theta) b_j)^2 = \theta^2 \sum_{j=1}^{n} a_j^2 + (1-\theta)^2 \sum_{j=1}^{n} b_j^2 + 2\theta(1-\theta) \sum_{j=1}^{n} a_j b_j.$$

Thus

$$\sum_{j=1}^{n} c_j^2 \leq \theta^2 + (1-\theta)^2 + 2\theta(1-\theta) \sum_{j=1}^{n} a_j b_j.$$

Next we prove that

$$\sum_{j=1}^{n} a_j b_j \leq 1.$$

Since

$$0 \leq \sum_{j=1}^{n} (a_j - b_j)^2 = \sum_{j=1}^{n} a_j^2 + \sum_{j=1}^{n} b_j^2 - 2 \sum_{j=1}^{n} a_j b_j$$

we have

$$0 \leq 2 - 2 \sum_{j=1}^{n} a_j b_j, \qquad \text{or} \qquad 1 - \sum_{j=1}^{n} a_j b_j \geq 0.$$

It follows that

$$\sum_{j=1}^{n} c_j^2 \leq \theta^2 + (1-\theta)^2 + 2\theta(1-\theta) = (\theta + (1-\theta))^2 = 1.$$

Problem 2.   Let $b > a$ and consider the function $f : \mathbb{R} \to \mathbb{R}$

$$f(x) = x^2$$

in this interval. Show that f is convex.

Solution 2.   We have to evaluate the difference

$$d := \theta f(x) + (1-\theta) f(y) - f(\theta x + (1-\theta) y).$$

We have to show that $d \geq 0$ for any $\mathbf{x}$ and $\mathbf{y}$ in C and any θ with $0 \leq \theta \leq 1$. From (1) we find

$$\begin{aligned}
d &= \theta x^2 + (1-\theta) y^2 - (\theta x + (1-\theta) y)^2 \\
&= \theta x^2 + (1-\theta) y^2 - \theta^2 x^2 - (1-\theta)^2 y^2 - 2\theta(1-\theta) xy \\
&= x^2 (\theta - \theta^2) + y^2 ((1-\theta) - (1-\theta)^2) - 2\theta(1-\theta) xy \\
&= x^2 \theta(1-\theta) + y^2 \theta(1-\theta) - 2\theta(1-\theta) xy \\
&= \theta(1-\theta)(x^2 + y^2 - 2xy).
\end{aligned}$$

Thus

$$d = \theta(1 - \theta)(x - y)^2.$$

Since $\theta \geq 0$, $1 - \theta \geq 0$ and $(x - y)^2 \geq 0$ it follows that $d \geq 0$.

Problem 3.   Let E^n be the n-dimensional Euclidean space. Let S be a closed convex set in E_n and $\mathbf{y} \notin S$.
(i) Show that there exists a unique point $\bar{\mathbf{x}} \in S$ with a minimum distance from $\mathbf{y}$.
(ii) Show that $\bar{\mathbf{x}}$ is the minimizing point if and only if $(\mathbf{x} - \bar{\mathbf{x}})^T (\bar{\mathbf{x}} - \mathbf{y}) \geq 0$ for all $\mathbf{x} \in S$.

Solution 3.   (i) Let

$$\inf \{ \, \|\mathbf{y} - \mathbf{x}\| : \mathbf{x} \in S \, \} = \gamma > 0.$$

There exists a sequence $\{\mathbf{x}_k\}$ in S such that $\|\mathbf{y} - \mathbf{x}_k\| \to \gamma$. We show that $\{\mathbf{x}_k\}$ has a limit $\bar{\mathbf{x}} \in S$ by showing that $\{\mathbf{x}_k\}$ is a *Cauchy sequence*. By the *parallelogram law*, we have

$$\|\mathbf{x}_k - \mathbf{x}_m\|^2 = 2\|\mathbf{x}_k - \mathbf{y}\|^2 + 2\|\mathbf{x}_m - \mathbf{y}\|^2 - \|\mathbf{x}_k + \mathbf{x}_m - 2\mathbf{y}\|^2.$$

It follows that

$$\|\mathbf{x}_k - \mathbf{x}_m\|^2 = 2\|\mathbf{x}_k - \mathbf{y}\|^2 + 2\|\mathbf{x}_m - \mathbf{y}\|^2 - 4 \left\| \frac{\mathbf{x}_k + \mathbf{x}_m}{2} - \mathbf{y} \right\|^2.$$

Note that $(\mathbf{x}_k + \mathbf{x}_m)/2 \in S$, and by definition of γ we have

$$\left\| \frac{\mathbf{x}_k + \mathbf{x}_m}{2} - \mathbf{y} \right\|^2 \geq \gamma^2.$$

Therefore

$$\|\mathbf{x}_k - \mathbf{x}_m\|^2 \leq 2\|\mathbf{x}_k - \mathbf{y}\|^2 + 2\|\mathbf{x}_m - \mathbf{y}\|^2 - 4\gamma^2.$$

By choosing k and m sufficiently large, $\|\mathbf{x}_k - \mathbf{y}\|^2$ and $\|\mathbf{x}_m - \mathbf{y}\|^2$ can be made sufficiently close to γ^2, hence $\|\mathbf{x}_k - \mathbf{x}_m\|^2$ can be made sufficiently close to zero. Therefore $\{\mathbf{x}_k\}$ is a Cauchy sequence and has a limit $\bar{\mathbf{x}}$. Since S is closed, $\bar{\mathbf{x}} \in S$. To show uniqueness, suppose that there is an $\bar{\mathbf{x}}' \in S$ such that

$$\|\mathbf{y} - \bar{\mathbf{x}}\| = \|\mathbf{y} - \bar{\mathbf{x}}'\| = \gamma.$$

As a result of the convexity of S,

$$\frac{1}{2}(\bar{\mathbf{x}} + \bar{\mathbf{x}}') \in S.$$

By the Schwarz inequality, we obtain

$$\left\| \mathbf{y} - \frac{\bar{\mathbf{x}} + \mathbf{x}'}{2} \right\| \leq \frac{1}{2} \|\mathbf{y} - \bar{\mathbf{x}}\| + \frac{1}{2} \|\mathbf{y} - \bar{\mathbf{x}}'\| = \gamma.$$

If strict inequality holds, we violate the definition of γ. Therefore equality holds, and we must have

$$\mathbf{y} - \bar{\mathbf{x}} = \lambda(\mathbf{y} - \bar{\mathbf{x}}')$$

for some λ. Since

$$\|\mathbf{y} - \bar{\mathbf{x}}\| = \|\mathbf{y} - \bar{\mathbf{x}}'\| = \gamma$$

it follows that $|\lambda| = 1$. Clearly, $\lambda \neq -1$, because otherwise $\mathbf{y} = (\bar{\mathbf{x}} + \bar{\mathbf{x}}')/2 \in S$, contradicting the assumption that $\mathbf{y} \notin S$. So $\lambda = 1$, $\bar{\mathbf{x}}' = \bar{\mathbf{x}}$, and uniqueness is established.

(ii) We need to show that

$$(\mathbf{x} - \bar{\mathbf{x}})^T (\bar{\mathbf{x}} - \mathbf{y}) \geq 0$$

for all $\mathbf{x} \in S$ is both a necessary and sufficient condition for $\bar{\mathbf{x}}$ to be the point in S closest to $\mathbf{y}$. To prove sufficiency, let $\mathbf{x} \in S$. Then,

$$\|\mathbf{y} - \mathbf{x}\|^2 = \|\mathbf{y} - \bar{\mathbf{x}} + \bar{\mathbf{x}} - \mathbf{x}\|^2 = \|\mathbf{y} - \bar{\mathbf{x}}\|^2 + \|\bar{\mathbf{x}} - \mathbf{x}\|^2 + 2(\bar{\mathbf{x}} - \mathbf{x})^T (\mathbf{y} - \bar{\mathbf{x}}).$$

Since $\|\bar{\mathbf{x}} - \mathbf{x}\|^2 \geq 0$ and $(\bar{\mathbf{x}} - \mathbf{x})^T (\mathbf{y} - \bar{\mathbf{x}}) \geq 0$ by assumption,

$$\|\mathbf{y} - \mathbf{x}\|^2 \geq \|\mathbf{y} - \bar{\mathbf{x}}\|^2$$

and $\bar{\mathbf{x}}$ is the minimizing point. Conversely, assume that $\|\mathbf{y} - \mathbf{x}\|^2 \geq \|\mathbf{y} - \bar{\mathbf{x}}\|^2$ for all $\mathbf{x} \in S$. Let $\mathbf{x} \in S$ and note that $\bar{\mathbf{x}} + \lambda(\mathbf{x} - \bar{\mathbf{x}}) \in S$ for $\lambda > 0$ and sufficiently small. Therefore,

$$\|\mathbf{y} - \bar{\mathbf{x}} - \lambda(\mathbf{x} - \bar{\mathbf{x}})\|^2 \geq \|\mathbf{y} - \bar{\mathbf{x}}\|^2.$$

Also

$$\|\mathbf{y} - \bar{\mathbf{x}} - \lambda(\mathbf{x} - \bar{\mathbf{x}})\|^2 = \|\mathbf{y} - \bar{\mathbf{x}}\|^2 + \lambda^2 \|\mathbf{x} - \bar{\mathbf{x}}\|^2 + 2\lambda(\mathbf{x} - \bar{\mathbf{x}})^T (\bar{\mathbf{x}} - \mathbf{y}).$$

From these two equations we obtain

$$\lambda^2 \|\mathbf{x} - \bar{\mathbf{x}}\|^2 + 2\lambda(\mathbf{x} - \bar{\mathbf{x}})^T (\bar{\mathbf{x}} - \mathbf{y}) \geq 0$$

for all $\lambda > 0$ and sufficiently small. Dividing by $\lambda > 0$ and letting $\lambda \to 0$, the result follows.

Problem 4.   (i) Show that the intersection of convex sets is convex.

(ii) Show that the Cartesian product $I_1 \times \cdots \times I_n \subset \mathbb{R}^n$ of the intervals

$$I_j := \{\, x \in \mathbb{R} : a_j \le x \le b_j \,\}, \quad j = 1, 2, \ldots, n$$

is convex.

Solution 4. (i) Let S and T be two convex sets. Let a and b be in $S \cap T$. Then for any $0 \le \theta \le 1$, $\theta a + (1 - \theta)b$ is in S, since S is convex, and $\theta a + (1 - \theta)b$ is in T, since T is convex. Hence $(\theta a + (1 - \theta)b) \in S \cap T$. Therefore $S \cap T$ is convex.

(ii) Let $\mathbf{x} = (x_1, \ldots, x_n)$, $\mathbf{y} = (y_1, \ldots, y_n)$ with

$$a_j \le x_j \le b_j, \qquad a_j \le y_j \le b_j, \qquad j = 1, \ldots, n.$$

Let $0 \le \theta \le 1$. Then if $\mathbf{z} := \theta \mathbf{x} + (1 - \theta)\mathbf{y}$ we have

$$z_j = \theta x_j + (1 - \theta)y_j \le \theta b_j + (1 - \theta)b_j = b_j, \qquad j = 1, \ldots, n$$
$$z_j = \theta x_j + (1 - \theta)y_j \ge \theta a_j + (1 - \theta)a_j = a_j, \qquad j = 1, \ldots, n.$$

Hence

$$a_j \le z_j \le b_j, \qquad j = 1, \ldots, n$$

and $\mathbf{z} \in I_1 \times \cdots \times I_n$. Thus $I_1 \times \cdots \times I_n$ is convex.

Problem 5. Let $y = f(x)$ be a convex function. Assume that f is twice differentiable. Thus $f''(x) > 0$ for all $x \in \mathbb{R}$. The *Legendre transformation* of the function f is a new function g of a new variable p as a maximum with respect to x at the point $x(p)$. Now we define

$$g(p) := F(p, x(p)). \tag{1}$$

The point $x(p)$ is defined by the extremal condition $\partial F / \partial x = 0$ i.e., $f'(x) = p$. Since f is convex, the point $x(p)$ is unique.

Let $f(x) = x^2$. Show that

$$F(p, x) = px - x^2, \qquad x(p) = \frac{1}{2}p, \qquad g(p) = \frac{1}{4}p^2.$$

Solution 5. Obviously, f is convex, since $d^2 f / dx^2 = 2 > 0$ for all $x \in \mathbb{R}$. Now

$$F(p, x(p)) = px - x^2$$

and $\partial F / \partial x = p - 2x$ or $p = 2x$. Therefore

$$g(p) = F(p, x(p)) = p\frac{p}{2} - \left(\frac{p}{2}\right)^2 = \frac{1}{4}p^2.$$

Problem 6.   Let $X \subset \mathbb{R}$ be an interval. A function $\psi : X \to \mathbb{R}$ is *convex* if for all $x_1, x_2 \in X$ and numbers $\alpha_1, \alpha_2 \geq 0$ with $\alpha_1 + \alpha_2 = 1$,

$$\psi(\alpha_1 x_1 + \alpha_2 x_2) \leq \alpha_1 \psi(x_1) + \alpha_2 \psi(x_2) . \tag{1}$$

This means that every chord of the graph of ψ lies above the graph. Let $\psi : X \to \mathbb{R}$ be convex, let $x_1, x_2, \ldots, x_n \in X$, and let $\alpha_1, \alpha_2, \ldots, \alpha_n \geq 0$ satisfy $\sum_{j=1}^{n} \alpha_j = 1$. Show that (*Jensen's inequality*)

$$\psi \left(\sum_{i=1}^{n} \alpha_i x_i \right) \leq \sum_{i=1}^{n} \alpha_i \psi(x_i) . \tag{2}$$

Solution 6.   For $n \geq 3$ and $\alpha_n \neq 1$ we use the inductive step

$$\psi \left(\sum_{i=1}^{n} \alpha_i x_i \right) = \psi \left((1 - \alpha_n) \sum_{i=1}^{n-1} \alpha_i (1 - \alpha_n)^{-1} x_i + \alpha_n x_n \right)$$

$$\leq (1 - \alpha_n) \psi \left(\sum_{i=1}^{n-1} \alpha_i (1 - \alpha_n)^{-1} x_i \right) + \alpha_n \psi(x_n)$$

by (1). Thus (2) follows from the inequality for $n - 1$, since

$$\sum_{i=1}^{n-1} a_i (1 - \alpha_n)^{-1} = 1 .$$

Problem 7.   Let A be an $n \times n$ positive semidefinite matrix over $\mathbb{R}$. Show that the positive semidefinite quadratic form

$$f(\mathbf{x}) = \mathbf{x}^T A \mathbf{x}$$

is a convex function throughout $\mathbb{R}^n$.

Solution 7.   We define $\widetilde{\mathbf{x}} := \lambda \mathbf{x}_2 + (1 - \lambda) \mathbf{x}_1$ for any vectors $\mathbf{x}_1, \mathbf{x}_2 \in \mathbb{R}^n$ and $\lambda \in [0, 1]$. Thus we obtain

$$\widetilde{\mathbf{x}}^T A \widetilde{\mathbf{x}} = (\lambda \mathbf{x}_2 + (1 - \lambda)\mathbf{x}_1)^T A (\lambda \mathbf{x}_2 + (1 - \lambda)\mathbf{x}_1)$$

$$= (\mathbf{x}_1 + \lambda(\mathbf{x}_2 - \mathbf{x}_1))^T A (\mathbf{x}_1 + \lambda(\mathbf{x}_2 - \mathbf{x}_1))$$

$$= \mathbf{x}_1^T A \mathbf{x}_1 + 2\lambda(\mathbf{x}_2 - \mathbf{x}_1)^T A \mathbf{x}_1 + \lambda^2 (\mathbf{x}_2 - \mathbf{x}_1)^T A (\mathbf{x}_2 - \mathbf{x}_1) .$$

Since $\mathbf{x}^T A \mathbf{x} \geq 0$ for all $\mathbf{x}$ and $\lambda \in [0, 1]$, we have $\lambda^2 \mathbf{x}^T A \mathbf{x} \leq \lambda \mathbf{x}^T A \mathbf{x}$ for all $\mathbf{x} \in \mathbb{R}^n$. Thus we can write

$$\widetilde{\mathbf{x}}^T A \widetilde{\mathbf{x}} \leq \mathbf{x}_1^T A \mathbf{x}_1 + 2\lambda(\mathbf{x}_2 - \mathbf{x}_1)^T A \mathbf{x}_1 + \lambda(\mathbf{x}_2 - \mathbf{x}_1)^T A (\mathbf{x}_2 - \mathbf{x}_1)$$

$$= \mathbf{x}_1^T A \mathbf{x}_1 + \lambda(\mathbf{x}_2 - \mathbf{x}_1)^T A \mathbf{x}_1 + \lambda(\mathbf{x}_2 - \mathbf{x}_1)^T A \mathbf{x}_2$$

$$= \lambda \mathbf{x}_2^T A \mathbf{x}_2 + (1 - \lambda) \mathbf{x}_1^T A \mathbf{x}_1 .$$

Chapter 22

Optimization

Problem 1. A number of linear optimization problems may be cast in the form

$$\text{Min } E = c_1 x_1 + c_2 x_2 + \cdots + c_n x_n \equiv \sum_{j=1}^{n} c_j x_j$$

with linear constraints

$$\sum_{j=1}^{n} a_{ij} x_j (\leq, =, \geq) b_i, \qquad i = 1, 2, \ldots, m$$

$$x_j \geq 0, \qquad j = 1, 2, \ldots, n$$

where the coefficients c_j and a_{ij} are constants and the notation $(\leq, =, \geq)$ signifies that any of the three possibilities may hold in any constraint. This is the standard linear programming problem.

Find the minimum of

$$E = -5x_1 - 4x_2 - 6x_3$$

subject to

$$x_1 + x_2 + x_3 \leq 100$$
$$3x_1 + 2x_2 + 4x_3 \leq 210$$
$$3x_1 + 2x_2 \leq 150$$
$$x_1, x_2, x_3 \geq 0$$

Solution 1.   We apply a gradient method. We convert to equalities by introducing three nonnegative *slack variables* x_4, x_5, x_6.

$$x_1 + x_2 + x_3 + x_4 = 100$$
$$3x_1 + 2x_2 + 4x_3 + x_5 = 210 \tag{1}$$
$$3x_1 + 2x_2 + x_6 = 150$$

$$5x_1 + 4x_2 + 6x_3 + E = 0$$

where

$$x_1, x_2, x_3, x_4, x_5, x_6 \geq 0.$$

A basic feasible solution to this system is

$$x_4 = 100, \quad x_5 = 210, \quad x_6 = 150, \quad x_1 = 0, \quad x_2 = 0, \quad x_3 = 0.$$

The nonzero variables x_4, x_5, x_6 are the basic variables at this stage. Thus $E = 0$. Now, computing the gradient of E yields

$$\frac{\partial E}{\partial x_1} = -5, \qquad \frac{\partial E}{\partial x_2} = -4, \qquad \frac{\partial E}{\partial x_3} = -6$$

so that improvement in E can be obtained by increasing x_1, x_2, and/or x_3. Since the magnitude of the gradient in the x_3 direction is greatest we move in x_3-direction. We retain x_1 and x_2 as nonbasic (zero). Thus we obtain

$$x_3 + x_4 = 100, \qquad 4x_3 + x_5 = 210.$$

Either x_4 or x_5 must go to zero (since x_3 will be nonzero) while the other remains nonnegative. If x_4 goes to zero, $x_3 = 100$ and $x_5 = -190$, while if x_5 goes to zero $x_3 = 52.5$ and $x_4 = 47.5$. That is

$$x_3 = \min\left(\frac{100}{1}, \frac{210}{4}\right) = 52.5$$

and x_5 is to be eliminated (set to zero) as x_3 is introduced. The calculation is one of dividing the right-hand column by the coefficient of x_3 and comparing. The new basic variables, x_4, x_3, and x_6 each appear in only a single equation including (2). This is done by dividing the second equation by the coefficient of x_3, then multiplying it by the coefficient of x_3 in each of the other equations, and subtracting to obtain an equivalent set of equations. The second equation is chosen because it is the one in which x_5 appears. Thus we find

$$\frac{1}{4}x_1 + \frac{1}{2}x_2 + x_4 - \frac{1}{4}x_5 = 47\frac{1}{2}$$

$$\frac{3}{4}x_1 + \frac{1}{2}x_2 + x_3 + \frac{1}{4}x_5 = 52\frac{1}{2} \tag{3}$$

$$3x_1 + 2x_2 + x_6 = 150$$

and

$$\frac{1}{2}x_1 + x_2 - \frac{3}{2}x_5 + E = -315. \tag{4}$$

The basic feasible solution is

$$x_1 = 0, \qquad x_2 = 0, \qquad x_5 = 0, \qquad x_4 = 47\frac{1}{2}, \qquad x_3 = 52\frac{1}{2}, \qquad x_6 = 150.$$

From (4) we obtain

$$E = -315 - \frac{1}{2}x_1 - x_2 + \frac{3}{2}x_5$$

so that the value of E in the basic variables is -315. Repeating the above procedure, we now find that the largest positive coefficient in (4) is that of x_2, and so x_2 is to be the new basic (nonzero) variable. The variable to be eliminated is found from the coefficients of the basic variable in system (3) in the form

$$x_2 = \min \left(\frac{47\frac{1}{2}}{\frac{1}{2}}, \frac{52\frac{1}{2}}{\frac{1}{2}}, \frac{150}{2} \right) = 75$$

which corresponds to eliminating x_6. Thus, we now use the Gauss-Jordan procedure to obtain x_2 in the third equation only. The result is

$$-\frac{1}{2}x_1 + x_4 - \frac{1}{4}x_5 - \frac{1}{4}x_6 = 10$$

$$x_3 + \frac{1}{4}x_5 - \frac{1}{4}x_6 = 15$$

$$\frac{3}{2}x_1 + x_2 + \frac{1}{2}x_6 = 75$$

$$-x_1 - \frac{3}{2}x_5 - \frac{1}{2}x_6 + E = -390$$

The basic feasible solution is then

$$x_1 = 0, \qquad x_5 = 0, \qquad x_6 = 0, \qquad x_4 = 10, \qquad x_3 = 15, \qquad x_2 = 75.$$

The corresponding value of E is $E = -390$. There are no positive coefficients in this system, and so this is the minimum.

Problem 2.   Let S be a nonempty convex set in $\mathbb{R}^n$.

Definition. A vector $\mathbf{x} \in S$ is called an *extreme point* of S if

$$\mathbf{x} = \lambda \mathbf{x}_1 + (1 - \lambda)\mathbf{x}_2$$

with $\mathbf{x}_1, \mathbf{x}_2 \in S$, and $\lambda \in (0, 1)$ implies $\mathbf{x} = \mathbf{x}_1 = \mathbf{x}_2$.

Definition. Let S be a nonempty convex set in $\mathbb{R}^n$ and let $f : S \to \mathbb{R}$ be convex. Then $\mathbf{y}$ is called a *subgradient* of f at $\bar{\mathbf{x}}$ if

$$f(\mathbf{x}) \geq f(\bar{\mathbf{x}}) + \mathbf{y}^T(\mathbf{x} - \bar{\mathbf{x}})$$

for all $\mathbf{x} \in S$.

Minimize

$$f(x_1, x_2) = \left(x_1 - \frac{3}{2}\right)^2 + (x_2 - 5)^2$$

subject to the four inequalities

$$-x_1 + x_2 \leq 2, \qquad 2x_1 + 3x_2 \leq 11, \qquad x_1 \geq 0, \qquad x_2 \geq 0.$$

Solution 2. A convex polyhedral set S is represented by the four inequalities. The extreme points of S are $(0,0)$, $(0,2)$, $(1,3)$, $(11/2,0)$. The function $f(x_1, x_2)$ is a convex function, which gives the square of the distance from the point $(3/2, 5)$.

Let $f : \mathbb{R}^n \to \mathbb{R}$ be a convex function, and S be a nonempty convex set in $\mathbb{R}^n$. Consider the problem to minimize f subject to $\mathbf{x} \in S$. The point $\bar{\mathbf{x}} \in S$ is an optimal solution if and only if f has a subgradient $\mathbf{y}$ at $\bar{\mathbf{x}}$ such that

$$\mathbf{y}^T(\mathbf{x} - \bar{\mathbf{x}}) \geq 0$$

for all $\mathbf{x} \in S$. We have

$$\nabla f(x_1, x_2) = (2(x_1 - 3/2), 2(x_2 - 5))^T,$$

where ^{T} denotes transpose. Thus the gradient vector of f at the point $(1, 3)$ is

$$\nabla f(1, 3) = (-1, -4)^T.$$

We see geometrically that the vector $(-1, -4)$ makes an angle of $\leq \pi/4$ with each vector of the form $(x_1 - 1, x_2 - 3)$, where $(x_1, x_2) \in S$. Thus the optimality condition given above is verified. Now, suppose that it is claimed that $(0,0)$ is an optimal point. Note that

$$\nabla f(0, 0) = (-3, -10)^T$$

and for each nonzero $\mathbf{x} \in S$, we have $-3x_1 - 10x_2 < 0$. Hence, the origin could not be an optimal point. Moreover, we can improve f by moving from $\mathbf{0}$ in the direction $\mathbf{x} - \mathbf{0}$ for any $\mathbf{x} \in S$. In this case, the best local direction is $-\nabla f(0,0)$, that is the direction $(3, 10)$. Thus the solution is $(x_1, x_2) = (1, 3)$.

Problem 3.   Many optimization problems also include inequalities. The *Karush-Kuhn-Tucker conditions* extend the Lagrange multiplier method to include inequality constraints. Given an optimization problem with convex domain $\Omega \subseteq \mathbb{R}^n$,

$$
\begin{aligned}
&\text{minimize} && f(\mathbf{x}), && \mathbf{x} \in \Omega \\
&\text{subject to} && g_k(\mathbf{x}) \geq 0 , && k = 1, \ldots, K \\
& && h_m(\mathbf{x}) = 0 , && j = 1, \ldots, M .
\end{aligned}
$$

For this problem we can construct the *Lagrange function*

$$
L(\mathbf{x}, \boldsymbol{\lambda}, \boldsymbol{\mu}) = f(\mathbf{x}) - \sum_{k=1}^{K} \lambda_k g_k(\mathbf{x}) - \sum_{m=1}^{M} \mu_m h_m(\mathbf{x})
$$

where λ_k and μ_m are the Lagrange multipliers. If the problem has a solution

$$
\mathbf{x}^* = (x_1^*, x_2^*, \ldots, x_n^*)
$$

i.e., $\min_\mathbf{x} f(\mathbf{x}) = f(\mathbf{x}^*)$ and all constraints are satisfied, then the following Karush-Kuhn-Tucker conditions hold

$$
\nabla f(\mathbf{x}^*) - \sum_{k=1}^{K} \lambda_k^* \nabla g_k(\mathbf{x}^*) - \sum_{m=1}^{M} \mu_m^* \nabla h_m(\mathbf{x}^*) = \mathbf{0}
$$

and

$$
\begin{aligned}
g_k(\mathbf{x}^*) &\geq 0 & k &= 1, 2, \ldots, K \\
h_m(\mathbf{x}^*) &= 0 & m &= 1, 2, \ldots, M \\
\lambda_k^* g_k(\mathbf{x}^*) &= 0 & k &= 1, 2, \ldots, K \\
\lambda_k^* &\geq 0 & k &= 1, 2, \ldots, K .
\end{aligned}
$$

In convex programming problems, the Karush-Kuhn-Tucker conditions are necessary and sufficient for a global minimum.

Solve the optimization problem given in problem 2 using the Karush-Kuhn-Tucker method. The constraints $-x_1 + x_2 \leq 2$, $2x_1 + 3x_2 \leq 11$ we have to rewrite as

$$
2 + x_1 - x_2 \geq 0, \qquad 11 - 2x_1 - 3x_2 \geq 0 .
$$

Solution 3.   The domain is convex and thus we can apply the Karush-Kuhn-Tucker condition. The Lagrangian is

$$
L(\mathbf{x}, \boldsymbol{\lambda}) = f(\mathbf{x}) - \lambda_1 x_1 - \lambda_2 x_2 - \lambda_3(2 + x_1 - x_2) - \lambda_4(11 - 2x_1 - 3x_2) .
$$

Thus we obtain the conditions

$$\frac{\partial L}{\partial x_1} = 0 \Rightarrow 2(x_1 - 3/2) - \lambda_1 - \lambda_3 + 2\lambda_4 = 0$$

$$\frac{\partial L}{\partial x_2} = 0 \Rightarrow 2(x_2 - 5) - \lambda_2 + \lambda_3 + 3\lambda_4 = 0$$

$$(2 + x_1 - x_2) \geq 0$$

$$(11 - 2x_1 - 3x_2) \geq 0$$

$$\lambda_3(2 + x_1 - x_2) = 0$$

$$\lambda_4(11 - 2x_1 - 3x_2) = 0.$$

and

$$x_1 \geq 0, \ x_2 \geq 0, \ \lambda_1 x_1 = 0, \ \lambda_2 x_2 = 0, \ \lambda_1 \geq 0, \ \lambda_2 \geq 0, \ \lambda_3 \geq 0, \ \lambda_4 \geq 0.$$

The solution is $x_1 = 1$, $x_2 = 3$, $\lambda_1 = 0$, $\lambda_2 = 0$, $\lambda_3 = 1$, $\lambda_4 = 1$.

Problem 4.   Use the Karush-Kuhn-Tucker to solve the following optimization problem. Minimize

$$f(\mathbf{x}) = (x_1 - 4)^2 + (x_2 - 4)^2$$

subject to the constraints $x_1 \geq 0$, $x_2 \geq 0$,

$$4 - x_1 - x_2 \geq 0, \qquad 9 - x_1 - 3x_2 \geq 0.$$

Solution 4.   The domain is convex and thus we can apply the Karush-Kuhn-Tucker condition. The Lagrangian is

$$L(\mathbf{x}, \lambda_1, \lambda_2, \lambda_3, \lambda_4) = f(\mathbf{x}) - \lambda_1 x_1 - \lambda_2 x_2 - \lambda_3(4 - x_1 - x_2) - \lambda_4(9 - x_1 - 3x_2).$$

Thus we obtain the conditions

$$\frac{\partial L}{\partial x_1} = 0 \Rightarrow 2(x_1 - 4) - \lambda_1 + \lambda_3 + \lambda_4 = 0$$

$$\frac{\partial L}{\partial x_2} = 0 \Rightarrow 2(x_2 - 4) - \lambda_2 + \lambda_3 + 3\lambda_4 = 0$$

$$(4 - x_1 - x_2) \geq 0$$

$$(9 - x_1 - 3x_2) \geq 0$$

$$\lambda_3(4 - x_1 - x_2) = 0$$

$$\lambda_4(9 - x_1 - 3x_2) = 0$$

and

$$x_1 \geq 0, \ x_2 \geq 0, \ \lambda_1 x_1 = 0, \ \lambda_2 x_2 = 0, \ \lambda_1 \geq 0, \ \lambda_2 \geq 0, \ \lambda_3 \geq 0, \ \lambda_4 \geq 0.$$

The solution is $x_1 = 2$, $x_2 = 2$, $\lambda_1 = 0$, $\lambda_2 = 0$, $\lambda_3 = 4$, $\lambda_4 = 0$.

Bibliography

Barbeau, E. J., *Polynomials*, Springer-Verlag, New York (1989).

Berger, M., Pansu, P., Berry, J. P. and Saint-Raymond, X., *Problems in Geometry*, Springer, New York (1984).

Bernstein, D. S., *Matrix Mathematics: Theory, Facts, and Formulas with Application to Linear Systems Theory*, Princeton University Press (2005).

Constantinescu, F. and Magyari, E., *Problems in Quantum Mechanics*, Pergamon Press, Oxford (1971).

Cronin, J. A., Greenberg, D. F. and Telegdi, V. L., *Graduate Problems in Physics*, Addison Wesley, Reading (1967).

de Souza, P. N. and Silva J.-N., *Berkeley Problems in Mathematics*, Springer, New York (1998).

Flügge, S., *Practical Quantum Mechanics*, Springer, Berlin (1974).

Gelbaum, B., *Problems in Analysis*, Springer, New York (1982).

Hardy, Y., Kiat Shi Tan and Steeb W.-H., *Computer Algebra with SymbolicC++*, World Scientific, Singapore (2008).

Howson A. G., *A Handbook of terms used in algebra and analysis*, Cambridge University Press (1972).

Knoop, K., *Problem Book in the Theory of Functions*, Volume I, Dover, New York (1952).

Krzyz, J. G., *Problems in Complex Variables Theory* Elsevier, New York (1971).

Larson, L. C., *Problem Solving Through Problems*, Springer, New York (1983).

Louisell W. H., *Quantum Statistical Properties of Radiation*, Wiley, New York (1973).

Polyanin A. D. and Zaitsev V. F., *Handbook of Nonlinear Partial Differential Equations*, Chapman and Hall/CRC (2003).

Rassias, John M., *Counter Examples in Differential Equations and Related Topics*, World Scientific, Singapore (1991).

Spiegel, M. R., *Advanced Calculus*, Schaum's Outline Series, McGraw Hill, New York (1974).

Spiegel, M. R., *Finite Differences and Difference Equations*, Schaum's Outline Series, McGraw Hill, New York (1971).

Spiegel, M. R., *Complex Variables*, Schaum's Outline Series, McGraw Hill, New York (1971).

Steeb, W.-H., *Matrix Calculus and Kronecker Product with Applications and C++ Programs*, World Scientific, Singapore (1997).

Steeb, W.-H., *Problems and Solutions in Introductory and Advanced Matrix Calculus*, World Scientific, Singapore (2006).

Steeb, W.-H. and Hardy Y. *Problems and Solutions in Quantum Computing and Quantum Information*, World Scientific, Singapore (2006).

Steeb, W.-H., *The Nonlinear Workbook*, 4th edition, World Scientific, Singapore (2008).

Steeb, W.-H., *Continuous Symmetries, Lie Algebras Differential Equations and Computer Algebra*, World Scientific, Singapore (2007).

Thorpe, J. A., *Elementary Topics in Differential Geometry*, Springer, New York (1979).

Tomescu, I., *Problems in Combinatorics and Graph Theory*, Wiley, New York (1985).

Zeidler E., *Quantum Field Theory II: Quantum Electrodynamics: A Bridge between Mathematicians and Physicists*, Springer, Heidelberg (2008).

Index